GW00384844

Modern Transmission Line
Theory and Applications

MODERN TRANSMISSION LINE THEORY AND APPLICATIONS

Lawrence N. Dworsky

Manager, Florida Network Research Laboratory
Communications Products Division
Motorola, Inc.

A Wiley-Interscience Publication

JOHN WILEY & SONS
New York · Chichester · Brisbane · Toronto

中華民國七十四年　月初版

發行所：國　興　出　版　社
發行人：楊　國　藩
新竹市西門街262號
電話：(035) 223129號
行政院新聞局局版台業字第0465號
總經銷：黎　明　書　店
新竹市中正路７２號
電話：(035) 229418號
郵政劃撥0070273—8號

實價：　200　元

Library of Congress Cataloging in Publication Data

Dworsky, Lawrence N　　　1943–
　Modern transmission line theory and applications.

　"A Wiley-Interscience publication."
　Includes bibliographies and index.
　1. Microwave transmission lines. 2. Microwave integrated circuits. 3. Electronic circuit design–Data processing. I. Title.

TK7876.D86　　　621.381'32　　　79-9082
ISBN 0-471-04086-X

Printed in the United States of America

10 9 8 7 6 5 4 3 2 1

For Gillian

Preface

The study of transmission line theory in electrical engineering curricula is usually limited to considerations of the circuit properties of lengths of transmission line with given properties. Unfortunately, this is no longer adequate. There is a continuing effort in the electronics industry to build smaller circuits that operate at a higher frequency, or faster, than last year's version. Microwave integrated circuit technology design techniques are tending more and more toward stripline or microstrip designs on high dielectric constant ceramic substrates. The result of this trend, insofar as the circuit designer is concerned, is that almost every interconnection in a circuit will exhibit transmission line properties. An immediate corollary is that if circuits are to be well designed, the transmission lines in the circuit must be appropriately treated as part of the circuit. It is no longer possible to separate the transmission line user from the transmission line designer—not only are the lines present, but their properties are functions of the circuit layout itself.

The purpose of this book is to extend the initial treatment of transmission line theory received by most electrical engineers to the point where transmission line effects can be properly considered, and transmission line properties can be calculated as a function of materials and geometries. Properties of stripline and microstrip circuits are emphasized, since these are the two line types that emerge naturally in the microwave integrated circuit–printed circuit layout.

This book is intended for students and engineers who have had some exposure to transmission line theory. Although the properties of transmission lines are derived from basic considerations, these derivations are brief—being intended as a review—and the underlying justifications are assumed to be understood. That is, it is assumed that the reader is familiar with the concepts of distributed circuits, wave propagation, and the constant interplay between field variables and circuit variables that takes place in descriptions of distributed circuits.

Once the circuit designer has learned to treat the transmission lines in the circuit properly, it becomes possible to take advantage of the different circuit functions that are realizable in single or coupled transmission line form. These functions include directional couplers, coupled line filters, and tapered line impedance transformers. Since microwave integrated circuits are fabricated using

vii

some form of printed circuit technology, circuit functions obtained by using the transmission lines that are printed along with the connecting lines are very economical and repeatable, hence desirable.

An important tool of the transmission line designer is the digital computer. The only practical way to design or analyze arbitrary transmission line geometries is by means of some numerical procedure using a computer. Therefore numerical approaches to transmission line analysis are treated in great detail, with many examples. On the other hand, no attempt has been made to treat exhaustively the number of different numerical approaches available. Instead, several approaches were followed through many examples so that the reader can see these approaches being applied to varying types of problems. In this way there is a reasonable chance that some example resembles an actual problem at hand and that a solution procedure, though possibly not the optimum one, can be found for most problems.

At the end of each chapter is an annotated reading list. The comments should help the interested reader to locate quickly a reference providing more detail on a topic of interest than is contained on these pages.

I thank my wife, Tamara, and my colleague, Dr. Melvyn Slater, for their help in structuring and reviewing the pages that follow. Without this help I could not have completed the job.

LAWRENCE N. DWORSKY

Fort Lauderdale
August 1979

Contents

Modern Transmission Line
Theory and Applications

1

The Transmission Line Equations

The electrical transmission line is an example of a one-dimensional propagating electromagnetic wave system. The equations governing this system may be derived from Maxwell's equations, either directly or from a circuit theory point of view. Although both these derivations lead to the same result it is instructive to examine them consecutively so that the two approaches can be compared.

1.1 THE TRANSMISSION LINE: A DEFINITION

A transmission line can be rigorously defined as any structure that guides a propagating electromagnetic wave from point a to point b. In other words, we can regard a transmission line as a set of boundary conditions to Maxwell's equations that allow the description of a one-dimensional propagating wave between two points.

The common use of the term "transmission line" is far more restrictive. It is usually required that the electrical length of the (transmission) line be at least several percent of a wavelength at the highest frequency of interest. Also, wave guides are excluded. That is, we require that the line propagate a signal at all frequencies from the frequency of interest down to *and including* dc, with the line characteristics varying in a smooth and continuous manner over this frequency range.

The statement above requires further discussion. At dc, an ideal (lossless) transmission line is surrounded by electric and magnetic fields that are normal both to each other and to the direction of energy propagation. This is the common transverse electromagnetic (TEM) mode of propagation. This is not to say, however, that a transmission line must propagate a signal in the TEM mode. As is shown below, there are several types of transmission line that cannot support TEM waves at frequencies other than zero. This situation arises

1

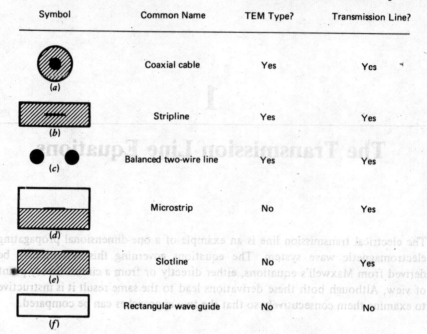

Symbol	Common Name	TEM Type?	Transmission Line?
(a)	Coaxial cable	Yes	Yes
(b)	Stripline	Yes	Yes
(c)	Balanced two-wire line	Yes	Yes
(d)	Microstrip	No	Yes
(e)	Slotline	No	Yes
(f)	Rectangular wave guide	No	No

Figure 1 Six common structures for one-dimensional wave propagation.

when there is an inhomogeneous dielectric cross section of the line normal to the direction of propagation. The transmission line definition given above is specifically constructed to include lines of these types while excluding conventional wave guides.

Figure 1 shows cross-sectional views of six types of one-dimensional wave guiding structures. The names usually associated with these (and many other) wave guiding structures came about through the first applications of these structures, and unfortunately often bear very little descriptive relation to the structures themselves.

1.2 THE TRANSMISSION LINE EQUATIONS
FROM MAXWELL'S EQUATIONS

To find a solution to Maxwell's equations specific enough to be compared with a set of circuit equations, a specific example must be chosen. This example may be any TEM system, or with certain approximations, any transmission line. As a simple example, consider the coaxial cable shown in Figure 2. In most practical cases the only propagating mode in coaxial cable is the TEM mode, and only this mode is considered here.

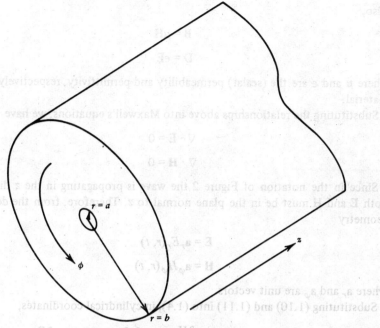

Figure 2 Coaxial cable cross section.

Assume that all of space is linear, isotropic, and homogeneous. Maxwell's equations are then

$$\nabla \cdot \mathbf{D} = \rho \tag{1.1}$$

$$\nabla \cdot \mathbf{B} = 0 \tag{1.2}$$

$$\nabla \times \mathbf{E} = - \frac{\partial \mathbf{B}}{\partial t} \tag{1.3}$$

$$\nabla \times \mathbf{H} = \mathbf{J} + \frac{\partial \mathbf{D}}{\partial t} \tag{1.4}$$

where **D** = electric displacement vector
   B = magnetic flux vector
   E = electric field intensity vector
   H = magnetic field intensity vector
   J = electric current density vector
   ρ = space charge density

Since the wave is propagating through a dielectric,

$$\rho = 0 \tag{1.5}$$

Also,

$$B = \mu H \tag{1.6}$$

$$D = \epsilon E \tag{1.7}$$

where μ and ϵ are the (scalar) permeability and permittivity, respectively, of the material.

Substituting the relationships above into Maxwell's equations, we have

$$\nabla \cdot E = 0 \tag{1.8}$$

$$\nabla \cdot H = 0 \tag{1.9}$$

Since in the notation of Figure 2 the wave is propagating in the z direction, both E and H must be in the plane normal to z. Therefore, from the described geometry

$$E = a_r E_r(r, t) \tag{1.10}$$

$$H = a_\varphi H_\varphi(r, t) \tag{1.11}$$

where a_r and a_φ are unit vectors.

Substituting (1.10) and (1.11) into (1.4), in cylindrical coordinates,

$$\nabla \times a_\varphi H_\varphi(r, t) = -a_r \frac{\partial H_\varphi}{\partial z} + a_z \frac{1}{r} \frac{\partial}{\partial r}(rH_\varphi) = a_r \epsilon \frac{\partial E_r}{\partial t} \tag{1.12}$$

For a TEM mode, the z-directed term in (1.12) must vanish. This is because $H_\varphi \sim 1/r$, therefore $\partial(rH_\varphi)/\partial r = 0$.

Equation 1.12 is now reduced to the scalar equation

$$\frac{\partial H_\varphi}{\partial z} = -\epsilon \frac{\partial E_r}{\partial t} \tag{1.13}$$

Integrating both sides of (1.13) about a circular path of radius r, where $a < r < b$, we have

$$\frac{\partial}{\partial z} \int_0^{2\pi} H_\varphi r \, d\varphi = -\frac{\partial}{\partial t} \int_0^{2\pi} \epsilon E_r r \, d\varphi \tag{1.14}$$

By Ampere's law the above then can be written as

$$\frac{\partial I}{\partial z} = -\frac{\partial}{\partial t} \int_0^{2\pi} \epsilon E_r r \, d\varphi \tag{1.15}$$

where I is the total current enclosed by the integration path—that is, the current flowing in the center conductor of the coaxial cable.

Multiplying and dividing the right-hand side of (1.15) by a length h, and using Gauss' law, yields

$$\frac{\partial I}{\partial z} = \frac{-1}{h} \frac{\partial}{\partial t} \int_0^{2\pi} \epsilon E_r rh \, d\varphi = -\frac{\partial}{\partial t} q_h \qquad (1.16)$$

where q_h is the charge per unit length enclosed by the cylindrical volume of length h and radius r. Since there is no free charge in the dielectric, this charge must reside on the center conductor.

Let us define the capacitance per unit length of the coaxial cable as C, where $C = q_h/V$. Since the electric field lines originating on the center conductor terminate on the outer conductor, V is the voltage between the inner and outer conductors, measured at any given z.

Rewriting (1.16) in terms of I, C, and V, we have

$$\frac{\partial I}{\partial z} = -C \frac{\partial V}{\partial t} \qquad (1.17)$$

Equation 1.17 is the first of the two transmission line equations. It couples the two variables I and V. A second equation is necessary to solve for these variables. This second transmission line equation is found by substituting (1.10) and (1.11) into (1.3). As above, in cylindrical coordinates,

$$\nabla \times a_r E_r = a_\varphi \frac{\partial E_r}{\partial z} - a_z \frac{1}{r} \frac{\partial E_r}{\partial \varphi} = -a_\varphi \mu \frac{\partial H_\varphi}{\partial t} = -a_\varphi \frac{\partial B_\varphi}{\partial t} \qquad (1.18)$$

The z-directed term in (1.18) must be zero from symmetry considerations. This leaves the scalar equation

$$\frac{\partial E_r}{\partial z} = -\frac{\partial B_\varphi}{\partial t} \qquad (1.19)$$

Integrating (1.19) along a radial path from $r = a$ to $r = b$, and at the same time multiplying and dividing the right-hand side by the length h, yields

$$\frac{\partial}{\partial z} \int_a^b E_r \, dr = \frac{-1}{h} \frac{\partial}{\partial t} \int_a^b hB_\varphi \, dr \qquad (1.20)$$

The left-hand side of (1.20) can be identified as $\partial V/\partial z$, while the right-hand side is $\partial \Phi_h/\partial t$, where Φ_h is the total magnetic flux per unit length passing through the rectangle of length h and width $b - a$. Defining the inductance per unit length of the cable as L, where $L = \Phi_h/I$, (1.20) becomes

$$\frac{\partial V}{\partial z} = -L \frac{\partial I}{\partial t} \qquad (1.21)$$

Equation 1.21, the second required transmission line equation, also couples the variables I and V. Equations 1.21 and 1.17, together with the appropriate boundary conditions, provide a complete description of the voltage and current waves as they propagate along the transmission line.

Before examining the same transmission line from a circuit theory viewpoint, let us consider some of the implications of the derivation above. If the inner and outer conductors of the coaxial cable are perfect conductors, no electric fields may exist within them. Therefore the transverse electric field E_r must exist only between the outer surface of the inner conductor and the inner surface of the outer conductor. In other words, E_r exists only in the dielectric region separating the two conductors.

Similarly, since the current flowing in the conductors exists only on the outer surface of the inner conductor and the inner surface of the outer conductor, the magnetic field H_φ must exist only in the dielectric region between the conductors.

Thus it has been said that when dealing with perfect conductors, current flows only along the surfaces of the conductors at which a magnetic field is present, and an electric field will exist only between the charged surfaces at which this electric field terminates. In the case of real metals, which are usually very good but not perfect conductors, the electric and magnetic fields will be shown to penetrate the surfaces slightly. Similarly, the currents will be shown to flow in a thin "skin" at and just below the same surfaces.

Poynting's vector, $\mathbf{E} \times \mathbf{H}$, predicts the power flow in the dielectric. Integrating Poynting's vector over the cross-sectional area of the dielectric yields the total power flow across any cross section of the line. This result must, of course, agree with the power flow calculation that is obtained from the voltage and current. The voltage-current calculation, however, does not bring out the point that the power is flowing in the dielectric—not in the conductors. Restating this observation, electric power does not flow in "wires"; rather, it flows in the fields surrounding the wires. As the coaxial cable example above demonstrated, the wires provide the boundary conditions for establishing a one-dimensional TEM wave solution to Maxwell's equations—that is, they guide the power flow.

1.3 THE TRANSMISSION LINE EQUATIONS FROM KIRCHHOFF'S EQUATIONS

The laws of circuit theory, Kirchhoff's voltage and current laws, can be used to derive the transmission line equations. To accomplish this, it is necessary to visualize a transmission line as a chain of discrete inductors and capacitors very closely spaced in a lattice structure. Figure 3 presents two such structures.

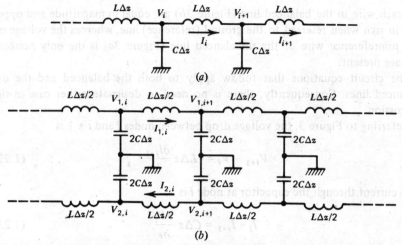

Figure 3 Balanced (b) and unbalanced (a) circuit models of a transmission line.

When a transmission line consists of two symmetrical conductors and a separate ground, or voltage reference, the transmission line is said to be balanced (see, e.g., Figure 1c). In this case if the pair of conductors has a total inductance per unit length of L, the inductance per unit length of each conductor must be $L/2$. If the capacitance per unit length between the conductors is C, to establish a ground reference symmetrically between the conductors, the capacitance per unit length to ground of each conductor is $2C$. Note that the separate ground often is not shown explicitly, as is the case in Figure 1c.

If the transmission line is represented by a chain of series inductors of value $L\Delta z/2$ and a chain of shunt capacitors of value $2C\Delta z$, the line can be approximated by the circuit of Figure 3b. The nodes are separated by the small distance Δz, and the node subscript i locates the node on the line according to $z = i(\Delta z)$.

In many cases it is convenient to consider one of the conductors of a transmission line as the voltage reference. This is particularly true when the geometry of the line causes the inductance per unit length to be much greater in one conductor than in the other. The conductor with the relatively small inductance is then taken as the voltage reference. When this is done, the total series inductance per unit length and the total shunt capacitance per unit length must be ascribed to the nonreference, or "ungrounded" conductor. Figure 3a depicts a transmission line, again as a chain of incremental elements. The most common example of this type of line is the coaxial cable. The outer conductor's inductance is much smaller than the inner conductor's inductance. The outer conductor is therefore chosen as the ground or reference line.

The two choices described above and shown in Figure 3 are often referred to as "balanced" and "unbalanced" transmission lines. This is because the voltages on each wire in the balanced line (Figure 3*b*) are equal in magnitude and opposite in sign when referred to the ground (reference) line, whereas the voltage on the nonreference wire of the unbalanced line (Figure 3*a*) is the only nonzero voltage present.

The circuit equations that follow apply to both the balanced and the unbalanced lines. Consequently, there is no need to designate either case in the discussion.

Referring to Figure 3, the voltage drop between nodes i and $i + 1$ is

$$V_{i+1} - V_i = -L\Delta z \frac{\partial I_{i+1}}{\partial t} \tag{1.22}$$

The current through the capacitor at node i is

$$I_i - I_{i+1} = C\Delta z \frac{\partial V_i}{\partial t} \tag{1.23}$$

Rearranging the two equations above into a more convenient form, we write

$$\frac{V_{i+1} - V_i}{\Delta z} = -L \frac{\partial I_{i+1}}{\partial t} \tag{1.24}$$

$$\frac{I_{i+1} - I_i}{\Delta z} = -C \frac{\partial V_i}{\partial t} \tag{1.25}$$

Assume that the increment Δz approaches zero. The left-hand side of both (1.24) and (1.25) would approach partial derivatives with respect to z because as Δz gets smaller, the $(i + 1)$th node is getting closer to the ith node. In the limit, (1.24) and (1.25) are identical to (1.17) and (1.21), namely,

$$\frac{\partial I}{\partial z} = -C \frac{\partial V}{\partial t} \tag{1.17}$$

$$\frac{\partial V}{\partial z} = -L \frac{\partial I}{\partial t} \tag{1.21}$$

When the ground return (voltage reference) path completely encloses the other wire(s) of a transmission line, the line is referred to as a shielded transmission line. In this case all electric and magnetic field lines terminate within the confines of the line cross section. Conversely, a balanced line can depend on the outside world for ground return paths. In this case, the line is unshielded and the field lines may extend an undeterminable distance from the line.

A distinction should be made at this point between an unshielded balanced line—even with ground returns only at infinity—and an antenna. Whereas a balanced, unshielded line has electric and magnetic fields that extend indefinitely, all power flow is along the line with no component normal to the line. This one-dimensional TEM wave is the proper transmission line wave. If, on the other hand, there were to exist a discontinuity or irregularity along the line, the fields may be disturbed in a manner that would cause radiation normal to the line. In this case the line would be acting, at least partially, as an antenna. It must be emphasized that this is a flaw condition, not the nature of the system.

In the case of the balanced line, a disturbance that might cause radiation could be either on the line or somewhere nearby—remember that the fields from an unshielded balanced line extend indefinitely. Since, in general, it is not possible to control the "neighborhood" through which a line must pass, a shielded line is usually preferable to avoid spurious radiation. On the other hand, since no dielectric is totally loss-free, when the "neighborhood" of a line can be controlled and line loss is intolerable, an open air, two-wire, balanced line is usually the optimum choice. As an example of this situation, consider the feed line to a transmitter on the top of a mountain, with the feed wires strung above the tree tops between towers erected specifically for this purpose.

1.4 LOW PASS FILTERS AND SIMULATED TRANSMISSION LINES

The cascade of series inductors and shunt capacitors treated in Section 1.3 can be redrawn as a cascade of identical T sections (Figure 4). For simplicity's sake, only the unbalanced line is shown. Again, the arguments pertain to both balanced and unbalanced lines. Each of the T sections (in Figure 4) can be recognized as a low pass filter section—that is, a two-port network that attenuates signals according to some monotonically increasing function of frequency. At first glance, it would seem natural to assume that a transmission line should have some sort of low pass filter characteristic.

This question can be investigated by finding the image impedance of one of the T sections. Image impedance is defined as that impedance which will appear at the input of a symmetric two-port network when the same impedance is used as the load to that network. Considering one of the T sections of Figure 4, the

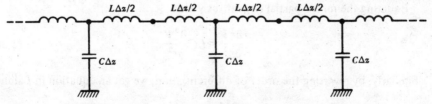

Figure 4 Cascade network of identical low pass T sections.

image impedance is calculated to be

$$Z_i(j\omega) = \sqrt{\frac{L}{C} - \left(\frac{\omega L \Delta z}{2}\right)^2} \qquad (1.26)$$

For Δz finite, Z_i is real only when ω is less than a "cutoff" frequency,

$$\omega_c = \frac{2}{\Delta z \sqrt{LC}} \qquad (1.27)$$

For $\omega > \omega_c$, Z_i is a pure imaginary number, and no power flows into (hence through) the T section.

On the other hand, as $\Delta z \to 0$, ω_c goes to infinity. This means that in the limit, Z_i is always real and independent of frequency. The value that Z_i approaches is defined to be Z_0, the characteristic impedance of the line. The (lossless) transmission line exhibits no low pass characteristics whatsoever.

On the other hand, suppose that we were to build the T sections shown in Figure 4 and study their two-port network properties over some frequency range where $\omega \ll \omega_c$. It can be shown that a properly designed T section can approximate the responses of a length of transmission line.

1.5 THE WAVE EQUATION

Equations 1.17 and 1.21 are coupled, first-order, partial differential equations in V and I. To look for solutions to them, it is useful to find an equation in V (or I) alone. Differentiating (1.17) with respect to t and (1.21) with respect to z, we have

$$\frac{\partial^2 I}{\partial z \partial t} = -C \frac{\partial^2 V}{\partial t^2} \qquad (1.28)$$

$$\frac{\partial^2 V}{\partial z^2} = -L \frac{\partial^2 I}{\partial t \partial z} \qquad (1.29)$$

Equating the mixed partial derivatives yields

$$\frac{\partial^2 V}{\partial z^2} = LC \frac{\partial^2 V}{\partial t^2} \qquad (1.30)$$

Similarly, by reversing the order of differentiation, we get an equation in I alone:

$$\frac{\partial^2 I}{\partial z^2} = LC \frac{\partial^2 I}{\partial t^2} \qquad (1.31)$$

Both V and I satisfy the same partial differential equation. Furthermore, the equation is the wave equation described in the last century by Maxwell for electromagnetic wave propagation.

Since the same partial differential equations (1.17 and 1.21) were also derived directly from Maxwell's equations, it should be possible to derive the wave equations directly from Maxwell's equations without going through the intermediate step of defining transmission line variables. Returning to the field equations, take the curl of both sides of (1.3) and substitute the result into (1.4):

$$\nabla \times (\nabla \times \mathbf{E}) = -\mu \nabla \times \frac{\partial \mathbf{H}}{\partial t} = -\mu \frac{\partial}{\partial t} (\nabla \times \mathbf{H}) = -\mu\epsilon \frac{\partial^2 \mathbf{E}}{\partial t^2} \qquad (1.32)$$

By vector identity,

$$\nabla \times \nabla \times \mathbf{E} = \nabla(\overset{0}{\cancel{\nabla \cdot \mathbf{E}}}) - \nabla^2 \mathbf{E} \qquad (1.33)$$

and therefore

$$\nabla^2 \mathbf{E} = \mu\epsilon \frac{\partial^2 \mathbf{E}}{\partial t^2} \qquad (1.34)$$

Equation 1.34 is a vector equation and must hold for each component of the vector. Referring to the coaxial cable example, $\mathbf{E} = \mathbf{a}_r E_r$ only. Therefore (1.34) simplifies to the scalar equation in E_r,

$$\frac{\partial^2 E_r}{\partial z^2} = \mu\epsilon \frac{\partial^2 E_r}{\partial t^2} \qquad (1.35)$$

Equation 1.35 has the same form as the equations in V and I, (1.30) and (1.31). Similarly, it can be shown that H_φ also satisfies the same wave equation. Physically, this is an expected result, since the voltage and current waves are defined directly in terms of the electric and magnetic fields and cannot move "independently" of them.

By direct substitution, (1.30), (1.31), and (1.35) all have solutions of the form

$$f\left(\frac{z}{v} \pm t\right) \qquad (1.36)$$

This solution represents a wave propagating in either direction along the z axis with a velocity v. From (1.30) or (1.31),

$$v^2 = \frac{1}{LC} \qquad (1.37)$$

and from (1.35),

$$v^2 = \frac{1}{\mu\epsilon} \qquad (1.38)$$

Since the electric field must be propagating at the same velocity as the voltage, the magnetic field, and the current, as was inferred above, we can write

$$LC = \mu\epsilon \qquad (1.39)$$

This equation is very useful when calculating transmission line parameters from the line geometry. It shows that L and C are not independent in a given dielectric material. In practice it is usually more convenient to calculate C than L, and (1.39) states that the calculation of C alone is sufficient in any given material.

1.6 LOSSY MEDIA

It is not possible to build totally loss-free transmission lines, since zero-resistance conductors and zero-conductance dielectrics (except for a vacuum) are simply not available. In many instances it is necessary to include the effects of conductor (ohmic) and dielectric losses in transmission line analyses and designs. It is possible to do this without invalidating (1.30) and (1.31) by introducing the concepts of complex inductance and complex capacitance in the following manner.

Consider an inductor L with its physical internal resistance represented by the series resistor R. The impedance of this series pair is

$$Z(j\omega) = j\omega L + R = j\omega \left[L - j\frac{R}{\omega} \right] \qquad (1.40)$$

The lossy inductor may be correctly represented in any (sinusoidal steady state) analysis that has been derived in terms of L by making the substitution

$$L \longrightarrow L - j\frac{R}{\omega} \qquad (1.41)$$

Similarly, consider a capacitor C, with its dielectric losses represented by the shunt conductance G. The admittance of this parallel pair is

$$Y(j\omega) = j\omega C + G = j\omega \left[C - j\frac{G}{\omega} \right] \qquad (1.42)$$

The lossy capacitor is then introduced to lossless analyses by the substitution

$$C \longrightarrow C - j\frac{G}{\omega} \qquad (1.43)$$

When dealing with dielectric materials, the properties of the dielectric usually are represented in terms of the relative permittivity and the dielectric loss factor.

These are defined in terms of the complex permittivity ϵ_c,

$$\epsilon_c = \epsilon_0 \left[\epsilon' - j\epsilon'' \right] \tag{1.44}$$

where ϵ_0 = permittivity of free space
ϵ' = relative permittivity (dielectric constant), often referred to as ϵ_r
ϵ'' = loss factor

These parameters can be related to material properties by rewriting (1.4) for the case of the sinusoidal steady state, and using Ohm's law,

$$\nabla \times \mathbf{H} = \mathbf{J} + j\omega\epsilon\mathbf{E} = \mathbf{E}\left[\sigma + j\omega\epsilon \right] = \mathbf{E}\left[j\omega\left(\epsilon - j\frac{\sigma}{\omega} \right) \right] \tag{1.45}$$

Comparing (1.45) to (1.44), we have

$$\epsilon = \epsilon_0 \epsilon' \tag{1.46}$$

$$\frac{\sigma}{\omega} = \epsilon_0 \epsilon'' \tag{1.47}$$

The ratio ϵ''/ϵ' is often written in terms of the loss tangent,

$$\tan \delta \equiv \frac{\epsilon''}{\epsilon'} \tag{1.48}$$

For most practical cases the loss tangent is a very small number, and

$$\delta \simeq \frac{\epsilon''}{\epsilon'} \tag{1.49}$$

It should be noted that the loss mechanisms discussed above are not the only possible ones. A dielectric material may have magnetic as well as dielectric loss mechanisms. The ferrite materials used in microwave circulators and isolators are a good example of this. Also, in a non-perfect conductor there is field penetration below the surface, and the possibility of dielectric loss exists. In practice, however, the latter case is not a very realistic consideration. The transmission line equations could be extended to consider the last two loss mechanisms. Since they are very unusual cases, however, they are not discussed any further.

The transmission line equations could have been derived explicitly considering loss mechanisms by including these mechanisms in the original cascaded T-section model (Figure 3). Without explicitly writing the difference equations and taking limits, the transmission line equations for the lossy lines are

$$\frac{\partial V}{\partial z} = -L\frac{\partial I}{\partial t} - RI \tag{1.50}$$

$$\frac{\partial I}{\partial z} = -C\frac{\partial V}{\partial t} - GV \tag{1.51}$$

Chapter 7 demonstrates that the field penetration occurring in nonperfect conductors causes a phase shift that can be represented by an inductance. The latter inductance is called the internal inductance of the transmission line. This is in contrast to the external inductance, which has been discussed exclusively thus far. The internal inductance represents only a very small fraction of the total (external + internal) inductance, and calculations of L and/or C can be carried out to a very high level of accuracy by considering the materials of the line to be lossless. When no reference to external or internal inductance is made, assume that the external inductance is being considered.

1.7 SUGGESTED FURTHER READING

1. W. C. Johnson, *Transmission Lines and Networks*, McGraw-Hill, New York, 1950. A very good introduction to transmission line theory, especially for someone with no previous exposure to the topic.

2. R. E. Collin, *Field Theory of Guided Waves*, McGraw-Hill, New York, 1960. A very thorough and rigorous treatise that discusses at length the mathematical nature of transmission line analysis. Probably the most authoritative book available on the subject, it is referenced frequently in the chapters that follow.

3. Sir Oliver Heaviside, *Electromagnetic Theory*, Dover, New York, 1956. This collection leads the reader through a fascinating series of articles written by the person most responsible for the development of transmission line theory. It also gives interesting insights into the nature of the developing electrical communications technology of late nineteenth century England, and the flowery nature of the technical debates of that time.

2

Time Domain Solutions to the Transmission Line Equations

The simplest solutions to the transmission line equations are those of the response of a line to a step of voltage or current at one end of the line. In studying these solutions, however, most of the important characteristics of transmission line analysis emerge. The basic concepts of characteristic impedance, reflection coefficient, and so on, are easily defined. In addition, two very useful electronic instruments, the charged line pulse generator and the time domain reflectometer, can be fully analyzed and understood in terms of these simple solutions.

2.1 TRAVELING WAVES ON A LOSSLESS LINE

Consider a uniform voltage wave solution to (1.30), as predicted by (1.36):

$$V(z, t) = f\left(\frac{z}{v} \pm t\right) \tag{2.1}$$

where the − sign indicates wave motion to the right (increasing z) and the + sign indicates wave motion to the left (decreasing z). The wave has a constant velocity $v = 1/\sqrt{\mu\epsilon}$.

Assuming that f is differentiable as necessary, we have

$$\frac{\partial V}{\partial z} = \frac{1}{v}\frac{\partial V}{\partial f} \tag{2.2}$$

$$\frac{\partial^2 V}{\partial z^2} = \frac{1}{v^2}\frac{\partial^2 V}{\partial f^2} \tag{2.3}$$

15

$$\frac{\partial V}{\partial t} = \frac{\partial V}{\partial f} \, (\mp 1) \tag{2.4}$$

$$\frac{\partial^2 V}{\partial t^2} = \frac{\partial^2 V}{\partial f^2} \tag{2.5}$$

Substituting (2.3) and (2.5) into (1.36), we have

$$\frac{1}{LC} \left(\frac{1}{v^2} \frac{\partial^2 V}{\partial f^2} \right) = \frac{\partial^2 V}{\partial f^2} \tag{2.6}$$

This equation confirms the assertion made previously [i.e., (1.35)].

Consider a semi-infinite line, extending from $z = 0$ to $z = \infty$. A wave launched at $z = 0$ will propagate in the $+z$ direction. There is no mechanism in this case for launching a wave in the $-z$ direction. Combining (1.21) and (2.2) yields

$$\frac{\partial I}{\partial t} = -\frac{1}{L} \frac{\partial V}{\partial z} = -\frac{1}{L} \left(\sqrt{LC} \frac{\partial V}{\partial f} \right) \tag{2.7}$$

and therefore for the wave traveling only in the $+z$ direction,

$$I(z, t) = + \sqrt{\frac{C}{L}} \, V(z, t) \tag{2.8}$$

The ratio of the voltage to the current at any point along the line is therefore a constant having the units of resistance. This resistance is defined as the *characteristic resistance* of the line. In anticipation of the sinusoidal steady state analysis of lossy lines, we lose no accuracy now by generalizing to the title of *characteristic Impedance* Z_0, defined in this case as

$$Z_0 = R_0 = \sqrt{\frac{L}{C}} \tag{2.9}$$

The input resistance at $z = 0$ is a real resistance of value Z_0. This means that any voltage or current source driving this line will be delivering power to a network that has been modeled using only inductors and capacitors. The apparent dilemma is resolved by remembering that the line under consideration is infinitely long. The propagating waves can never reach the "end" of the line. The flow of power into the line represents a continuous charging of the line.

In the general case of a line of finite length, it is of course possible to have waves traveling in both directions simultaneously. Since the system is linear, superposition holds, and the general solution to the transmission line equations can be written as follows:

$$V(z, t) = f(\sqrt{LC}z - t) + g(\sqrt{LC}z + t) \tag{2.10}$$

$$I(z, t) = \frac{1}{Z_0} \, [f(\sqrt{LC}z - t) - g(\sqrt{LC}z + t)] \tag{2.11}$$

The wave traveling to the left in the equations above, $g(z, t)$, is derived in the same manner as $f(z, t)$, noting that in the case of g, $v = -1/\sqrt{LC}$. Physically, this can be understood by picturing the voltage waves as adding directly at any point along the line, while the current waves are subtracting at any point along the line.

2.2 THE FINITE LENGTH LOSSLESS LINE

Consider a line of length h, characteristic impedance Z_0, and wave velocity v. The line is driven at $z = 0$ by a voltage source V_s, with internal resistance R_s, and terminated at $z = h$ with a resistance R_L. Figure 5 illustrates this circuit. We adopt the convention that a wave traveling to the right will have a subscript +, and a wave traveling to the left will have a subscript −. In general, then,

$$V(z, t) = V_+ + V_- \qquad (2.12)$$

$$I(z, t) = I_+ + I_- = \frac{V_+ - V_-}{Z_0} \qquad (2.13)$$

Noting that a current going to the right ($+z$) is taken as positive, by Ohm's law at $z = h$ we have

$$R_L = \frac{V(h, t)}{I(h, t)} \qquad (2.14)$$

for all t.

Substituting (2.12) and (2.13) into (2.14), we write

$$R_L = \frac{V_+(h, t) + V_-(h, t)}{\dfrac{V_+(h, t) - V_-(h, t)}{Z_0}} \qquad (2.15)$$

The ratio V_-/V_+, found by rearranging the equation above, is defined to be the voltage reflection coefficient, or simply the reflection coefficient Γ:

$$\Gamma \equiv \frac{V_-}{V_+} = \frac{R_L - Z_0}{R_L + Z_0} \qquad (2.16)$$

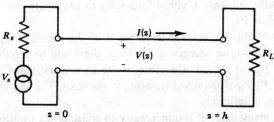

Figure 5   Finite length line driven by a real source and terminated in a real load.

Equation 2.16 reveals a great deal of information about transmission lines in electrical circuits. First, it predicts that if $R_L = Z_0$, then $\Gamma = 0$. This means that a wave going to the right, launched at $z = 0$, will be totally absorbed by R_L. Furthermore, at $z = 0$ this situation cannot be differentiated from that of the semi-infinite line discussed earlier. When $\Gamma = 0$ the line is said to be perfectly matched, properly matched, or simply matched at $z = h$.

Second, if $\Gamma \neq 0$, an incident wave (from the left) at $z = h$ *must* give rise to a reflected wave, originating at $z = h$, and traveling to the left. In this case the line is said to be improperly matched, or mismatched, at $z = h$.

Third, if R_L is passive (i.e., positive), the range of values that Γ may take on is $-1 \leqslant \Gamma \leqslant 1$. Conversely, if a measurement of Γ (as yet undescribed) showed $|\Gamma| > 1$, it could be concluded immediately that R_L was not positive.

Since the current wave as well as the voltage wave must be examined at $z = h$, it is equally important to consider a current reflection coefficient. Following the definition of Γ, let the current reflection coefficient be defined as

$$\frac{I_-}{I_+} = \left[\frac{-V_-}{Z_0}\right] \bigg/ \left[\frac{V_+}{Z_0}\right] = -\Gamma \qquad (2.17)$$

The current reflection coefficient is so readily defined in terms of Γ that it is pointless to name a new term for it.

Equations 2.14 to 2.17 are valid for all time. The voltage and current waves, of course, travel at a finite velocity. This means that for some period of time after the waves are launched at $z = 0$, V and I are identically 0 at $z = h$. Equations 2.16 and 2.17 are still satisfied during this time, but trivially.

A convenient means of picturing the propagation of a voltage (or current) step along a lossless line is the position-time diagram (Figure 6). In this diagram the horizontal axis represents position along the line and allows values from $z = 0$ to $z = h$. The vertical axis represents time, and allows values from $t = 0$ to $t = \infty$. A voltage wave front V_0, originating at $(0, 0)$, travels through the "space" of the diagram with a slope $\Delta t / \Delta z = 1/v$, and reaches $z = h$ at $t = T = h/v$.

Ahead of the voltage wave, for $t < T$, $V = 0$. Behind it, $V = V_0$. Let the reflection coefficient at $z = h$ be Γ_L, $\Gamma_L \neq 0$. At $t = T$ a wave $V_1 = \Gamma_L V_0$ originates at $z = h$ and travels back toward $z = 0$. This reflected wave reaches $z = 0$ at $t = 2T$. Ahead of this reflected wave ("ahead" meaning to the left in this case) $V = V_0$. Behind it, $V = V_0 + V_1$.

At $t = 2T$, the reflected wave reaches $z = 0$. If the source that launched the wave at $t = 0$ has a source resistance $R_s \neq Z_0$, there will be a reflection coefficient at $z = 0$, Γ_s. At $t = 2T$, therefore, if the source resistance $\neq Z_0$, a wave $V_2 = \Gamma_s V_1 = \Gamma_s \Gamma_L V_0$ is launched traveling to the right. This multiple reflection process continues indefinitely.

In practice, in many cases, it is unnecessary to consider the multiple reflections when one is interested only in the final, dc steady state, response. For example,

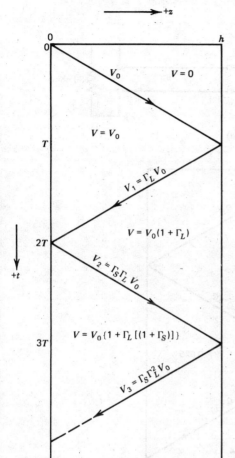

$+z$

0 h

V_0 $V = 0$

$V = V_0$

T

$V_1 = \Gamma_L V_0$

$V = V_0(1 + \Gamma_L)$

$2T$

$V_2 = \Gamma_S \Gamma_L V_0$

$V = V_0\{1 + \Gamma_L[(1 + \Gamma_S)]\}$

$3T$

$V_3 = \Gamma_S \Gamma_L^2 V_0$

$+t$

Figure 6 Basic position-time diagram.

turning on an automobile's headlights might be considered to be a case of a resistance at the far end of a transmission line, with a constant voltage wave front launched at the near end. Obviously under such conditions the multiple reflections, constituting a transient situation, quickly "relax" to a steady state. Consider the following example:

Example. A 1 volt battery with an internal resistance of 10 ohms is connected, at $t = 0$, to a 10 m length of lossless transmission line. This length of line is found to have an inductance of 01.0 mH and a capacitance of 0.4 μF. The line is terminated by a 30 ohm resistor. This and the accompanying position-time diagram appear in Figures 7a and 7b, respectively.

Since the transmission line parameters are expressed as L and C per unit

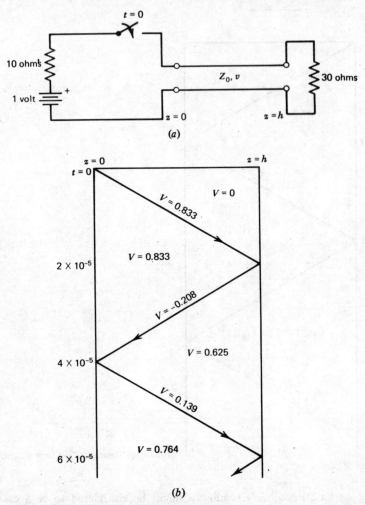

Figure 7 (a) Example circuit. (b) Corresponding position-time diagram.

length, from the data above we have

$$L = 10^{-4} \text{ H/m}$$

$$C = 4 \times 10^{-8} \text{ F/m}$$

The line therefore has a characteristic impedance of

$$Z_0 = \sqrt{\frac{L}{C}} = 50 \text{ ohms}$$

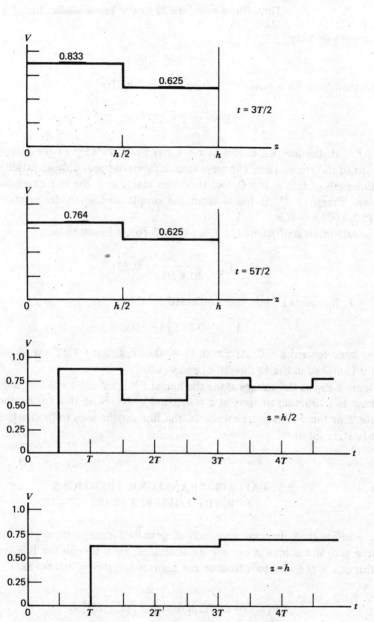

Figure 8 $V(t)$ for several values of z, example.

21

and a wave velocity

$$v = 5 \times 10^5 \text{ m/sec}$$

The transit time for a wave front along a 10 m length is

$$T = \frac{h}{v} = 2 \times 10^{-5} \text{ sec}$$

At $t = 0$, the lumped circuit at $z = 0$ has no knowledge of the length of the line, or of its termination. The wave launching circuit sees only an infinitesimally small length of line at $z = 0$, and therefore reacts as if the line extended indefinitely. Therefore V_0 is found from the simple voltage-divider relation $V_0 = (1)(50)/(60) = \frac{5}{6}$ volt.

The reflection coefficient at $z = h$, using (2.16), is found to be

$$\Gamma_L = \frac{30 - 50}{30 + 50} = -0.25$$

At $t = T$, V_0 reaches $z = h$, and a reflected wave

$$V_1 = -0.25 \left(\tfrac{5}{6}\right) = -0.208$$

starts back toward $z = 0$. At $z = 0$, $\Gamma_s = -0.667$, and at $t = 2T$ a wave of $+0.139$ volt is launched in the $+z$ direction, and so on.

Figure 8 shows the voltage along the line at $t = 1.5\,T$ and $t = 2.5\,T$, and also the voltage as a function of time at $z = h/2$ and $z = h$. Note that from either the V versus z or the V versus t viewpoints, the line can be seen to be charging to the steady state solution.

2.3 LAPLACE TRANSFORM SOLUTIONS
FOR THE LOSSLESS LINE

The position-time diagram approach as described cannot be applied when the source and/or the load is not a pure resistance, even though the line is lossless. In this case it is useful to introduce the Laplace transform, defined as

$$F(z, s) = \mathcal{L}\{f(z, t)\} = \int_0^\infty f(z, t)e^{-st}\, dt \qquad (2.18)$$

The inverse transform problem is not discussed here, and standard tables can be consulted when required. Notationally, the functional dependence of variables is shown explicitly when they are not obvious from context. In this way it is possible to avoid introducing a plethora of new variables. Also, the problem is simplified by assuming that all initial conditions are zero. The Laplace transformation can be pursued further by consulting the suggested readings at the end of this chapter.

Applying (2.18) to (1.17) and (1.21), we get the transmission line equations in the transform or "spectral" domain,

$$\frac{dV(z, s)}{dz} = - sLI(z, s) \tag{2.19}$$

$$\frac{dI(z, s)}{dz} = -sCV(z, s) \tag{2.20}$$

These equations are ordinary differential equations in V and I. Differentiating (2.19) and then substituting the result into (2.20), we have

$$\frac{d^2 V}{dz^2} = -sL(-sCV) = s^2 LCV = \gamma^2 V \tag{2.21}$$

where $\gamma = s\sqrt{LC}$.

Similarly, solving for I yields

$$\frac{d^2 I}{dz^2} = \gamma^2 I \tag{2.22}$$

The general solutions to the equations above are

$$V(z, s) = Ae^{-\gamma z} + Be^{\gamma z} \tag{2.23}$$

and

$$I(z, s) = \frac{1}{Z_0} [Ae^{-\gamma z} - Be^{\gamma z}] \tag{2.24}$$

Consider first the example given in the last section, as shown in Figure 8. The boundary conditions for this example are that

$$V(0, s) = V_0(s) - Z_s I(0, s) \tag{2.25}$$

and

$$V(h, s) = +Z_L I(h, s) \tag{2.26}$$

where $V_0(s)$ = the (transformed) source voltage.

Substituting these conditions into the general solutions and solving for A and B, we have

$$A = \frac{V_0 Z_0 (Z_L + Z_0) e^{\gamma h}}{e^{\gamma h}(Z_s + Z_0)(Z_L + Z_0) - e^{-\gamma h}(Z_L - Z_0)(Z_s - Z_0)} \tag{2.27}$$

$$B = \frac{V_0 Z_0 (Z_L - Z_0) e^{\gamma h}}{e^{\gamma h}(Z_s + Z_0)(Z_L + Z_0) - e^{-\gamma h}(Z_L - Z_0)(Z_s - Z_0)} \tag{2.28}$$

from which

$$V(z, s) = V_0 Z_0 \left[\frac{(Z_L + Z_0)e^{-\gamma(z-h)} + (Z_L - Z_0)e^{+\gamma(z-h)}}{e^{\gamma h}(Z_s + Z_0)(Z_L + Z_0) - e^{-\gamma h}(Z_L - Z_0)(Z_s - Z_0)} \right]$$

(2.29)

Rewriting this solution in terms of the reflection coefficients,

$$V(z, s) = \frac{V_0 Z_0 e^{-\gamma z}}{Z_s + Z_0} \left[\frac{1 + \Gamma_L e^{-2\gamma(z-h)}}{1 - \Gamma_L \Gamma_s e^{-2\gamma h}} \right]$$

(2.30)

The steady state solution can be found without first finding the inverse transform of $V(z, s)$ by using the final value theorem of Laplace transform theory,

$$\lim_{t \to \infty} f(z, t) = \lim_{s \to 0} s F(z, s)$$

(2.31)

Applying this theorem to (2.29) and (2.30), we find

$$\lim_{t \to \infty} V(z, t) = \lim_{s \to 0} \frac{s V_0(s) Z_L}{Z_L + Z_s} = \lim_{s \to 0} \frac{s V_0(s) Z_0}{Z_s + Z_0} \frac{1 + \Gamma_L}{1 - \Gamma_L \Gamma_s}$$

(2.32)

As is expected, the result is a function neither of z nor of Z_0. In the case of the dc source (battery) switched on at $t = 0$, $V_0(s) = V_{dc}/s$ (the Laplace transform of a voltage step of height V_{dc}), and the steady state solution is

$$V(z) = V_{dc} \frac{Z_L}{Z_L + Z_s}$$

(2.33)

The time required for a given transient to approach the steady state solution with some measure of accuracy is of course a function of all the parameters involved [see (2.30)]. It should be kept in mind, however, that lumped circuit theory is based on the premise that connecting "wires" are lossless and short enough so that the time required for the steady state solutions to be reached on all transmission lines (wires) is much shorter than any other time of interest.

An interesting paradox arising from this discussion is the case of the ideal voltage source ($R_s = 0$) connected to a lossless line that is terminated in an open circuit. In this case $\Gamma_s = -1$ and $\Gamma_L = +1$. From (2.32) we find that $V(z) = V_{dc}$, the voltage of the source. Intuition and practical experience lead us to the same conclusion. The same situation can be described using a position-time diagram, as in Figure 9. The diagram predicts a voltage at $z = 0$ of V_{dc}, a necessary identity, but at $z = h$ the voltage is seen to be a square wave in time, alternating between 0 and $2V_{dc}$ volts, and never settling to any steady state value at all. The resolution of this discrepancy is left to the interested reader as an exercise.

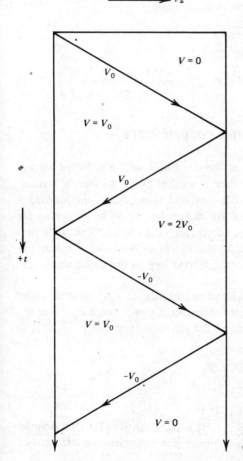

Figure 9 Proposed paradox, position-time diagram.

The Laplace transform solution technique can be extended directly to problems with inductive and/or capacitive sources and loads by merely extending the definition of the reflection coefficient to include the Laplace transform impedance as Z_s or Z_L. For example, consider the circuit of Figure 8, but with the load consisting of a series resistor and inductor. In this case $Z_L = R_L + sL$. Assume that the line is matched at the source, that is, $R_s = Z_0$. From (2.29), taking the source to be a unit step of 1 volt at $t = 0$, and recalling the definition of γ,

$$V(z, s) = \frac{1}{2}\left[\frac{e^{-z/v}}{s} + \left(\frac{(R_L - Z_0)/(R_L + Z_0)}{s} + \frac{2Z_0/(R_L + Z_0)}{s + (R_L + Z_0)/L}\right)e^{-(2T-z/v)s}\right]$$

$$(2.34)$$

Referring to a table of inverse transforms, we can write

$$V(z,t) = \frac{u(t-z/v)}{2} + \frac{1}{2}\left[\frac{R_L - Z_0}{R_L + Z_0} + \frac{2Z_0}{R_{L_1} + Z_0}\right.$$

$$\left.\cdot \exp\left(\frac{-(R_L + Z_0)}{L}\left(t - 2T + \frac{z}{v}\right)\right)\right] u\left(t - 2T + \frac{z}{v}\right) \quad (2.35)$$

2.4 TRANSMISSION COEFFICIENTS

The reflection coefficient is the ratio of the amplitude of the reflected voltage wave to that of the incident voltage wave at a point along the line. If a transmission line were to be terminated with a second transmission line, having a characteristic impedance different from its own, there would be no reason for the reflection coefficient concept not to remain valid. In addition, it is perfectly reasonable to define a transmission coefficient as the ratio of the amplitude of the voltage wave launched into the second line to the amplitude of the voltage wave incident at the junction.

Consider a transmission line of characteristic impedance Z_{01}, terminated by one end of a second transmission line, of characteristic impedance Z_{02}. Assume that there is a voltage wave V_1 incident at the junction from line 1. The reflection coefficient, as seen by line 1, is

$$\Gamma = \frac{Z_{02} - Z_{01}}{Z_{02} + Z_{01}} \quad (2.36)$$

and the amplitude of the reflected wave is $V_1 \Gamma$.

The total voltage at the junction is $V_1(1 + \Gamma)$. The quantity $(1 + \Gamma)$ is defined as the voltage transmission coefficient, or simply the transmission coefficient of the junction.

At the transmission line junction described above, the current incident from line 1, I_1, causes a reflected current wave of value $-\Gamma I_1$. To satisfy Kirchhoff's current law at the junction, there must be a current wave launched into line 2 of amplitude $(1 - \Gamma)I_1$. Multiplying the transmitted current by the transmitted voltage, we find that the fraction of the incident power that is transmitted into line 2 is

$$\frac{P_{\text{transmitted}}}{P_{\text{incident}}} = (1 + \Gamma)(1 - \Gamma) = 1 - \Gamma^2 \quad (2.37)$$

The fraction of the incident power that is reflected back into line 1 is of course

$$\frac{P_{\text{reflected}}}{P_{\text{incident}}} = \Gamma^2 \quad (2.38)$$

These two power ratios usually are measured in decibels and referred to as the transmission loss and return loss, respectively, at the junction. That is,

$$\text{transmission loss} = 10 \, \text{Log}_{10}(1 - \Gamma^2) \quad (2.39)$$

$$\text{return loss} = 10 \, \text{Log}_{10}(\Gamma^2) = 20 \, \text{Log}_{10}(\Gamma) \quad (2.40)$$

It is, of course, perfectly valid to use the foregoing definitions when measuring power transmitted into and returned from the junction of a transmission line and a lumped termination, or even when referring to the junction of two lumped element networks. The latter case appeared often in the "image parameter" technique of filter design, but is not common in the recent literature.

2.5 THE CHARGED LINE PULSE GENERATOR

A low loss transmission line charged to a uniform voltage can be used as the basis for a generator of very low rise time and fall time pulses of predetermined duration. The operation of this pulse generator is best described by using an easily proved network theorem, the *switch closing theorem*.

Consider a linear passive network N, with two ports that are connected, respectively, to (1) an impedance Z, through which a current $I(s)$ is flowing, and (2) an open switch. Assume that inside the network there are sources and/or initial conditions such that while $I(s)$ is flowing through Z, a voltage $V(s)$ appears across the (open) switch terminals.

If the switch were to be replaced by a voltage source $V(s)$, no current would flow through this source. Therefore closing the switch must be equivalent to adding a second source $-V(s)$, in series with the first source. At Z, $I(s)$ would change to $I(s) + \Delta I$ because of this new source. Since N is linear, the new source—or its equivalent, closing the switch—must be the cause of the current increment ΔI.

An analogous theorem, the switch opening theorem, is described in the following manner. A closed switch is replaced by a current source equal in amplitude to the current flowing through the switch. The switch opening is then equivalent to adding a second current source, equal in amplitude and opposite in sign to the first source, in parallel with the first source.

Consider a length of lossless transmission line charged uniformly to a voltage V_0. The line is connected, through an open switch, to a resistor whose resistance is equal to the characteristic impedance of the line, $R = Z_0$. Figure 10a shows this circuit. Referring to the switch closing theorem, the circuit voltages and currents are not changed by replacing the open switch with the voltage source V_0 as shown in Figure 10b. At $t = 0$, assume that the switch in the original circuit is closed. Using the switch closing theorem, this is equivalent to adding a second voltage source, of value $-V_0$, in series with the first source, as shown in Figure

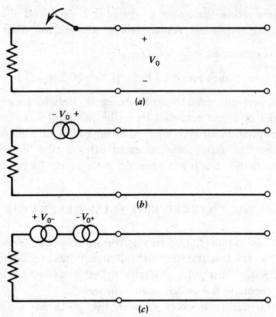

Figure 10 Development of the application of the switch closing theorem to the charged line pulse generator.

10c. By superposition, the current through R can be found by considering only the second voltage source connected in the circuit with the transmission line having zero initial conditions (i.e., initially uncharged).

Figure 11 gives the position-time diagram for this circuit and the accompanying graph for the voltage across R. Here a pulse of voltage $V_0/2$ appears across R for a time $2h/v = 2h\sqrt{LC}$. The rise time of the pulse is determined principally by the speed at which the switch closes, and the fall time of the pulse is determined primarily by the quality of the transmission line.

Figure 12 presents a common practical circut for a charged line pulse generator: the charging supply is conneeted permanently through a large resistor R_c. Although this arrangement limits the generator to operating at a very low duty cycle, it is convenient and does not noticeably affect the pulse forming characteristic of the line.

Originally, charged line pulse generators were built using mercury-wetted reed relays as switches. A mercury-wetted reed relay is capable of extremely fast, clean switching. Pulse generators with rise times as low as 0.06 nsec have been built using these relays. The principal drawback of a relay is that the mechanically moving, liquid-wetted relay contacts cannot be relied on to "fire" at an

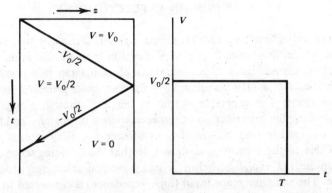

Figure 11 Position-time diagram for the charged line pulse generator.

exact time after the relay coil is energized. One solution to this problem is to trigger all necessary measurement circuitry with the generator output, then to delay the signal to be measured long enough for the measurement circuitry to respond to the trigger. The required delay can be achieved by simply passing the pulse through a suitable length of transmission line.

In recent years, fast switching semiconductor devices have replaced the reed relay in all but the most critical low rise time situations. The solid state semiconductor switch can be triggered reliably and also will run at a much higher pulse repetition rate than the mechanical relay.

The quality of the transmission line used for pulse generation is the principal parameter in determining the maximum duration of the generated pulses. A lossy line will cause decay of stored energy as the pulse is forming, with a resulting deterioration in both the level and the fall time of the generated pulse.

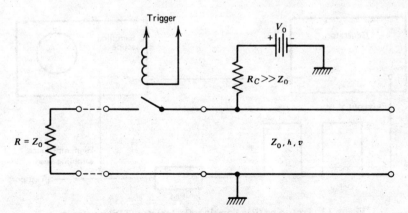

Figure 12 Practical circuit for a charged line pulse generator.

2.6 TIME DOMAIN REFLECTOMETRY

Time domain reflectometry (TDR) is a measurement technique that is used to study transmission line systems. It is very useful in locating and evaluating discontinuities on a line. Figure 13 is a functional block diagram of a possible TDR system. The heart of a TDR system is a low rise time pulse generator, running in a repetitive mode. The generator output is sampled through a large resistor (thereby avoiding the introduction of unnecessary line mismatch), and this sampled output is used to trigger a sampling oscilloscope. A delay cable is placed in the output line of the generator, as shown, so that the sampling scope can view the entire pulse. As explained below, a pad (matched attenuator) follows the delay cable. The sampling scope input (high impedance) is connected to the output of the pad. The pad output is also the test port of the measurement system.

As an example of the measurement capabilities of this TDR system, consider the problem of a length of transmission line having a suspected discontinuity at some unknown location along its 5 m length. Assume that the line has a characteristic impedance of 50 ohms, as does the transmission line and pad of the TDR system. The propagation velocity along the suspected line is also unknown.

Assume that the generator in the TDR system has been adjusted to produce pulses having a 5 nsec duration and an output level of 1 volt when driven into a Z_0 termination. The pulse repetition rate is very low (less than 10^4/sec). The length of line to be analyzed is connected to the TDR output port and is shorted at its far end. From the sampling scope display of the TDR system (Figure 14), the following information can be seen.

Pulse 1 is the "outgoing" pulse leaving the TDR system. Its known height and duration serve as calibration references for the display. Pulse 2 is caused by a reflection somewhere along the line. Its amplitude is identically Γ, the reflection

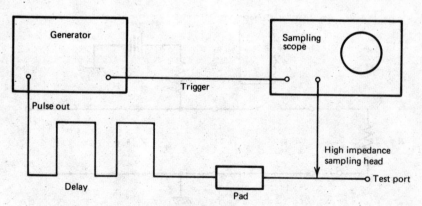

Figure 13 Prototype time domain reflectometry (TDR) system.

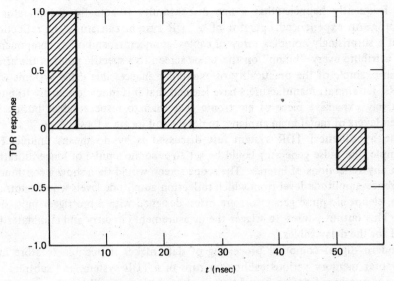

Figure 14 Sampling scope display for the TDR example.

coefficient at the point of reflection. Solving (2.16) for R_L, we find that

$$R_L = 50 \frac{1 + 0.5}{1 - 0.5} = 150 \text{ ohms}$$

This means that in a 50 ohm line, there is some type of discontinuity with an effective series resistance of 100 ohms.

Note that pulse 2 is traveling back into the TDR system. It is absorbed without reflection by the pad, travels along the pulse forming cable, is reflected back along the pulse forming cable in some unspecified manner, and is further attenuated by the pad before reappearing at the sampling scope probe. Therefore a larger pad attenuation always results in less loss of resolution due to multiple reflections in the system. Unfortunately, a correspondingly larger voltage output is required of the pulse generator to produce a useable signal level in this case.

Pulse 3 is the pulse that was transmitted past the discontinuity, reflected by the short at the far end of the line, again transmitted past the discontinuity and finally returned to the TDR port. The time of arrival of pulse 3 is 50 nsec after originally leaving the TDR port, therefore the transit time along the line must be 25 nsec. Since the line is 5 m long, $v = 5/(25 \times 10^{-9}) = 2 \times 10^8$ m/sec. This velocity is two-thirds that of a wave in free space, therefore $\epsilon_r = (\frac{3}{2})^2 = 2.25$ for the dielectric of the line. Since pulse 2 returned after 20 nsec, the discontinuity must be $\frac{10}{25}$ of the length of the line away from the TDR port, which is 0.4(5) = 2 m from the TDR port.

It frequently happens that a line discontinuity is actually a connector or adaptor. An experienced operator of a TDR system can study the reflections from a surprisingly complex array of cables, connectors, and other components, and attribute every "bump" on the scope screen to a specific point in the array. As an example of the practicality of examining inaccessible cable systems with TDRs, jet aircraft manufacturers have learned that it is much less costly to buy a relatively expensive piece of electronic gear than to inspect cables buried between layers of metal in an airplane, to detect and locate a flaw.

The hypothetical TDR system just discussed is by no means unique. For example, the pulse generator could be set to generate a pulse of longer duration than any reflections of interest. The scope screen would then show a continuous reference amplitude level from which reflection amplitude levels would subtract. Also, electronic pulse generators are often designed with a pretrigger pulse output. This output is used to trigger the measurement circuitry and eliminates the need for the delay cables.

Modern digital computer processing of data makes it possible to store in a computer memory various calibration runs of a TDR system, and subtract the relevant numbers from an actual data run. In this manner the accuracy and the resolution of a test instrument are enhanced greatly. This approach eliminates the need for extremely costly, precise transmission line connectors and cables.

2.7 SUGGESTED FURTHER READING

1. Gustav Doetsch (translated), *Guide to the Application of Laplace Transforms*, Van Nostrand, Princeton, N.J., 1961. A very complete and thorough set of tables of transforms, and a well written presentation of Laplace transform theory.

2. Ernst Weber, *Linear Transient Analysis*, Vol. II, *Two-Terminal Pair Networks and Transmission Lines*, Wiley, New York, 1956. A very authoritative text on the details of transform analysis as applied to transmission line theory, and the necessary mathematics.

3

The Alternating Current
Steady State

The form of the solutions to the transmission line equations in the sinusoidal steady state has already been suggested by the form of the Laplace transform solution presented in Chapter 2. In terms of the sinusoidal steady state solutions it is sometimes convenient to consider a length of line as a two-port network, converting line parameters to the parameters of some convenient two-port network set. This is perhaps most useful when designing a general purpose computer network analysis program. As with lumped circuit elements, the parameter set chosen is a matter of utility in dealing with the problem at hand.

When dealing with reflections on lines in the sinusoidal steady state, the standing wave patterns along the line lead to the concept of voltage standing wave ratio (VSWR). The relationship between VSWR and reflection coefficient leads to what is probably the most useful engineering design aid ever conceived, the Smith chart.

3.1 GENERAL SOLUTIONS

Considering the Laplace transform solutions to the transmission line equations, and making the usual substitution, $s \rightarrow j\omega$, we have

$$V(z) = A_1 e^{-\gamma z} + A_2 e^{\gamma z} \tag{3.1}$$

$$I(z) = \frac{1}{Z_0} (A_1 e^{-\gamma z} - A_2 e^{\gamma z}) \tag{3.2}$$

where A_1 and A_2 are the magnitudes of the voltage waves traveling in the $+z$ and $-z$ directions, respectively.

Now Z_0 and γ are functions of ω. Replacing L and C by the general terms for

33

a lossy line as suggested in Chapter 1, we write

$$Z_0 = \sqrt{\frac{R + j\omega L}{G + j\omega C}} \tag{3.3}$$

$$\gamma = \sqrt{(R + j\omega L)(G + j\omega C)} \tag{3.4}$$

It is conventional in transmission line literature to define the parameters Z and Y as

$$Z = R + j\omega L \tag{3.5}$$

$$Y = G + j\omega C \tag{3.6}$$

In terms of Z and Y, this gives

$$Z_0 = \sqrt{\frac{Z}{Y}} \tag{3.7}$$

$$\gamma = \sqrt{ZY} \tag{3.8}$$

In general, a lossy line is dispersive. That is, a nonsinusoidal wave shape propagating along the line will change shape. Short rise and fall times (of pulses), representing high order harmonic terms in a Fourier expansion of the wave shape, usually attenuate faster than low order terms, and a pulse will "smear out." This has been understood empirically since the early days of telegraphy, when operators knew that in rainy weather they had to slow down their Morse code transmissions to maintain intelligibility.

Consider a single-frequency sinusoid propagating on an infinitely long line, or on a properly matched line of finite length. Since in this case there is no reflected wave, assuming that the wave was launched at $z = 0$, A_2 in (3.1) and (3.2) must be zero. Writing γ in terms of its real and imaginary components, let

$$\gamma = \alpha + j\beta \tag{3.9}$$

Equations 3.1 and 3.2 become

$$V(z) = V_0 e^{-\alpha z} e^{-j\beta z} \tag{3.10}$$

$$I(z) = \frac{V_0}{Z_0} e^{-\alpha z} e^{-j\beta z} \tag{3.11}$$

where $V_0 = V(0)$.

In this case, obviously, $V(z) = Z_0 I(z)$. Note that Z_0 is no longer a simple resistance. Equations 3.10 and 3.11 show that the magnitudes of V and I are both decaying with increasing z. For $V_0 = V(0)$ and $I_0 = I(0)$, the power being delivered to the line at $z = 0$ is

$$P_{in} = \frac{V_0 I_0}{2} \cos(\theta) \qquad (3.12)$$

where θ is the angle between V_0 and I_0. At some point $z = h$,

$$V(h) = V_0 e^{-\alpha h} e^{-j\beta h} \qquad (3.13)$$

$$I(h) = I_0 e^{-\alpha h} e^{-j\beta h} \qquad (3.14)$$

Since both V and I have undergone the same phase shift between $z = 0$ and $z = h$, the angle between them, θ, has not changed. The power flow crossing the point $z = h$ is therefore

$$P(h) = \frac{V_0 I_0}{2} e^{-2\alpha h} \cos(\theta) \qquad (3.15)$$

The attenuation of the line is

$$att = -10 \, Log_{10} \left[\frac{P(h)}{P_{in}} \right] = -10 \, Log_{10}(e^{-2\alpha h}) = 8.69\alpha h \qquad dB \qquad (3.16)$$

Dividing both sides of (3.16) by h, the attenuation per unit length is

$$\frac{att}{h} = 8.69\alpha \qquad (3.17)$$

Since α is the real part of γ, it is clear from (3.14) that α is an unwieldy function of the line parameters. For the practical case of a low loss line, the following approximations can be made.

If $R \ll \omega L$ and $G \ll \omega C$, then Z_0 can be taken to be its essentially lossless (pure resistive) value; $V(z)$ and $I(z)$ are then in phase, and

$$P(z) = \frac{V_0 I_0}{2} e^{-2\alpha z} \qquad (3.18)$$

In this case the rate of decrease of P with respect to z is

$$\frac{\partial P}{\partial z} = -2\alpha P \qquad (3.19)$$

This rate of change must be equal to the power absorbed by the line per unit length (a function of z). For the low loss line this can be estimated as the sum of the conductor losses (due to R) and the dielectric losses (due go G):

$$2\alpha P(z) = \frac{I^2(z)R}{2} + \frac{V^2(z)G}{2} = \frac{V^2(G + R/Z_0^2)}{2} \qquad (3.20)$$

Noting that $P(z) = [V(z)I(z)]/2 = [V^2(z)]/2Z_0$, we write

$$\alpha = \frac{GZ_0 + R/Z_0}{2} \tag{3.21}$$

Combining the result above with (3.17) gives

$$\frac{\text{att (dB)}}{h} = 4.35 \left(GZ_0 + \frac{R}{Z_0} \right) \tag{3.22}$$

The parameters R, G, and Z_0 in (3.22) are functions of frequency, line geometry, and materials. If these functions were known, it would be possible to design a line for minimum attenuation at a given frequency. In many situations functional dependences of these parameters are known in an approximate manner, and good if not optimum designs are possible.

3.2 THE FINITE LENGTH LINE AND ITS TWO-PORT NETWORK PARAMETERS

Chapter 1 showed that a finite length of transmission line, terminated in arbitrary resistances, would exhibit wave reflections at these resistances. This situation can be extended directly to the sinusoidal steady state. In this case Γ is defined as follows:

$$\Gamma = \frac{Z_L - Z_0}{Z_L + Z_0} \tag{3.23}$$

and is, in general, a complex number.

The finite length line is an example of a linear two-port network. Various transfer properties of interest (input impedance for an arbitrary termination, voltage gain, insertion loss, etc.) can be found using either of two general approaches. It is always possible to start with (3.1) and (3.2), and using the required definition, derive any relationship. On the other hand, a much more efficient approach is to start with (3.1) and (3.2) and derive the matrix elements for any convenient two-port parameter set. Since the relations between the various two-port parameter sets and the different transfer relations are standard handbook items, the latter derivation is much more general.

Figure 15 shows a length h of line, with the input and output voltage and current variables labeled in terms of two-port network conventions. Writing (3.1) and (3.2) at $z = 0$, we have

$$V(0) = V_1 = A_1 + A_2 \tag{3.24}$$

$$I(0) = I_1 = \frac{A_1 - A_2}{Z_0} \tag{3.25}$$

and at $z = h$

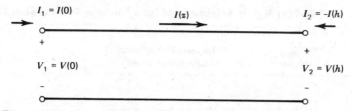

Figure 15 Two-port network parameter conventions for a length of line.

$$V(h) = V_2 = A_1 e^{-\gamma h} + A_2 e^{\gamma h} \tag{3.26}$$

$$I(h) = -I_2 = \frac{A_1 e^{-\gamma h} - A_2 e^{\gamma h}}{Z_0} \tag{3.27}$$

The choice of two-port parameter sets is arbitrary. Consider the Z parameters. These are defined by the equations

$$V_1 = Z_{11} I_1 + Z_{12} I_2 \tag{3.28}$$

$$V_2 = Z_{21} I_1 + Z_{22} I_2 \tag{3.29}$$

Solving for Z_{11},

$$Z_{11} = \left.\frac{V_1}{I_1}\right|_{I_2=0} = Z_0 \left[\frac{A_1 + A_2}{A_1 - A_2}\right]_{I_2=0} \tag{3.30}$$

The boundary condition $I_2 = 0$ yields $A_1 = A_2 e^{2\gamma h}$; therefore

$$Z_{11} = Z_0 \frac{e^{2\gamma h} + 1}{e^{2\gamma h} - 1} = Z_0 \coth(\gamma h) \tag{3.31}$$

Similarly,

$$Z_{12} = Z_{21} = Z_0 \operatorname{csch}(\gamma h) \tag{3.32}$$

and

$$Z_{22} = Z_{11} = Z_0 \coth(\gamma h) \tag{3.33}$$

Note that there are only two independent parameters. This is because a length of uniform transmission line is a symmetric reciprocal network. For that matter, whereas the symmetry property may be lost by going to a tapered (nonuniform) transmission line, the reciprocity property is never lost.

Each two-port parameter set has an accompanying lumped element equivalent circuit. For the Z parameters, this is a T network.

Table 1 shows the defining relationships for the more common two-port parameter sets, the equivalent network associated with each set, and the values of the parameters of each set for a length of uniform transmission line. Also, several useful transfer relations are given in terms of each of the parameter sets.

Table 1   Two-Port Matrix Relationships for a Uniform Transmission Line

Matrix Set	Defining Relations	Uniform Transmission Line Matrix Terms	Z_{in}	Z_{out}
$[Z_{ij}]$	$V_1 = Z_{11}I_1 + Z_{12}I_2$ $V_2 = Z_{21}I_1 + Z_{22}I_2$	$Z_0 \begin{bmatrix} \coth(\theta) & \operatorname{csch}(\theta) \\ \operatorname{csch}(\theta) & \coth(\theta) \end{bmatrix}$	$\dfrac{\Delta_z + Z_{11}Z_L}{Z_{22} + Z_L}$	$\dfrac{\Delta_z + Z_{22}Z_s}{Z_{11} + Z_s}$
$[Y_{ij}]$	$I_1 = Y_{11}V_1 + Y_{12}V_2$ $I_2 = Y_{21}V_1 + Y_{22}V_2$	$Y_0 \begin{bmatrix} \coth(\theta) & -\operatorname{csch}(\theta) \\ -\operatorname{csch}(\theta) & \coth(\theta) \end{bmatrix}$	$\dfrac{Y_{22} + Y_L}{\Delta_y + Y_{11}Y_L}$	$\dfrac{Y_{11} + Y_s}{\Delta_y + Y_{22}Y_s}$
$[h_{ij}]$	$V_1 = h_{11}I_1 + h_{12}V_2$ $I_2 = h_{21}I_1 + h_{22}V_2$	$\begin{bmatrix} Z_0 \tanh(\theta) & \operatorname{sech}(\theta) \\ -\operatorname{sech}(\theta) & Y_0 \tanh(\theta) \end{bmatrix}$	$\dfrac{\Delta_h + h_{11}Y_2}{h_{22} + Y_L}$	$\dfrac{h_{11} + Z_s}{\Delta_h + h_{22}Z_s}$
$\begin{bmatrix} AB \\ CD \end{bmatrix}$	$V_1 = AV_2 - BI_2$ $I_1 = CV_2 - DI_2$	$\begin{bmatrix} \cosh(\theta) & Z_0 \sinh(\theta) \\ Y_0 \sinh(\theta) & \cosh(\theta) \end{bmatrix}$	$\dfrac{AZ_L + B}{CZ_L + D}$	$\dfrac{DZ_s + B}{CZ_s + A}$

Notes:
1. $\theta \equiv \gamma h$.
2. Z_{in}, Z_0, A_V, are as shown:

$$A_V = V_2/V_1$$

3.3 STANDING WAVES, VSWR AND THE SMITH CHART

Consider a finite length of line that is mismatched at one end ($z = h$). In this case the voltage along the line is given by (3.1), and neither A_1 nor A_2 is zero.

Table 1 (*Cont.*)

A_v	Reciprocity Condition	Symmetry (Uniformity) Condition	Uniform Line Model
$\dfrac{Z_{21}Z_L}{\Delta_z + Z_{11}Z_L}$	$Z_{12}=Z_{21}$	$Z_{11}=Z_{22}$	Series $Z_{11}-Z_{12}$, $Z_{11}-Z_{12}$; shunt Z_{12}
$\dfrac{-Y_{21}}{Y_{22}+Y_L}$	$Y_{12}=Y_{21}$	$Y_{11}=Y_{22}$	Series $-Y_{12}$; shunt $Y_{11}+Y_{12}$, $Y_{11}+Y_{12}$
$\dfrac{-h_{21}}{\Delta_h + h_{11}Y_L}$	$h_{12}=-h_{21}$	$\Delta_h=1$	h_{11}; $h_{12}V_2$; $h_{21}I_1$; h_{22}
$\dfrac{Z_L}{B+AZ_L}$	$\Delta_{ABCD}=1$	$A=D$	

3. $Z_0 \equiv 1/Y_0$.

4. $\Delta_z \equiv \det\,[Z]$, etc.

Writing the time domain equation corresponding to the phasor equation (3.1),

$$V(z,t) = A_1 \cos(\omega t + j\gamma z) + A_2 \cos(\omega t - j\gamma z) \qquad (3.34)$$

and by trigonometric identity,

$$V(z,t) = (A_1 + A_2)\cos(\omega t)\cos(j\gamma z) - (A_1 - A_2)\sin(\omega t)\sin(j\gamma z)$$

$$(3.35)$$

In the case of a lossless line, γ is purely imaginary. The voltage and current wave forms along a lossless line are therefore periodic in z, with a period $\lambda = v/f$.

The length λ is defined to be "one wavelength." If a line is greater than 1 quarter-wavelength long at a given frequency, there must exist at least one point z_1, where $\cos (j\gamma z_1) = 0$, and another point z_2, a quarter-wavelength away from z_1, where $(j\gamma z_2) = 0$.

At z_2, the peak voltage is $A_1 + A_2$, and at z_1 the peak voltage is $A_1 - A_2$. In general, a graph of peak voltage versus z would show a sinusoid with a period λ varying between a peak of $|A_1| + |A_2|$ and a minimum of $|A_1| - |A_2|$. Since the locations of the peak and the minimum are stationary in time, the wave is called a "standing wave." The ratio of the peak amplitude to the minimum amplitude, the voltage standing wave ratio (VSWR), is defined as follows:

$$\text{VSWR} = \frac{|A_1| + |A_2|}{|A_1| - |A_2|} = \frac{1 + |A_2/A_1|}{1 - |A_2/A_1|} = \frac{1 + |\Gamma_L|}{1 - |\Gamma_L|} \tag{3.36}$$

A measurement of VSWR is usually made with a length of rigid coaxial cable, slotted so that a sliding probe can be inserted into the line and the electric field sampled. The voltage detected by the probe is rectified and fed to a voltmeter. Slotted line measurements determine wavelength from the distance between the voltage maximum and minimum, and VSWR from the same voltages. From (3.36), $|\Gamma_L|$ can be calculated once the VSWR is known. The angle of Γ_L can be inferred from the location of the first peak and minimum of voltage with respect to the end of a slotted line that is terminated by an unknown impedance. Since Γ_L is then completely specified, the unknown load impedance can be calculated from the definition of Γ_L.

Consider a polar plot of Γ, as in Figure 16a. Restricting the discussion to real passive load impedances, the polar plot need only extend from $|\Gamma| = 0$ to $|\Gamma| = 1$. For convenience, assume that we are dealing with a system in which $Z_0 = 1$. In this case,

$$\Gamma = \frac{Z_L - 1}{Z_L + 1} \tag{3.37}$$

Equation 3.37 can be considered to be a mapping function. That is, for every value of Z_L (with a real part greater than zero), (3.37) describes a $1:1$ correspondence between a point on the Z_L plane and a point on the Γ plane. For example $Z_L = 1$ maps into $\Gamma = 0$, the center of the circle. A short circuit, $Z_L = 0$, maps into $\Gamma = -1 + j0$, and so on.

Consider a length of line h, characteristic impedance $Z_0 = 1$, terminated by some impedance Z_L. The input impedance of this line can be found by di‗ding (3.1) by (3.2) and referencing to an origin at the load impedance:

$$Z(-h) = \frac{V(-h)}{I(-h)} = \frac{A_1 e^{\gamma h} + A_2 e^{-\gamma h}}{A_1 e^{\gamma h} - A_2 e^{-\gamma h}} = \frac{1 + (A_2/A_1)e^{-2\gamma h}}{1 - (A_2/A_1)e^{-2\gamma h}} \tag{3.38}$$

In this reference frame, A_2/A_1 is identically the reflection coefficient at the

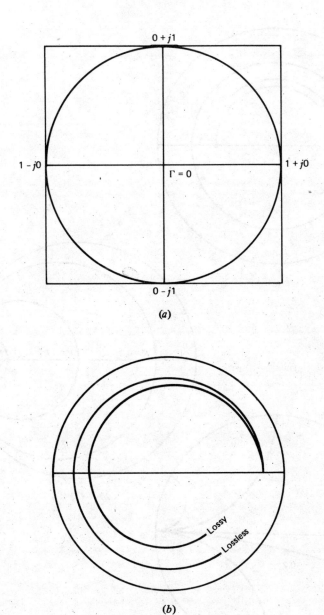

(a)

(b)

Figure 16 (a) Γ in polar coordinates. (b) A real terminating impedance reflected back to the input of a transmission line: lossless and lossy line cases. (c) Real terminating impedances reflected back to the input of half-wave and quarter-wave (lossless) lines. (d) Several curves of constant R superimposed on the Polar Γ curve. (e) Several curves of constant X superimposed·on the polar Γ curve.

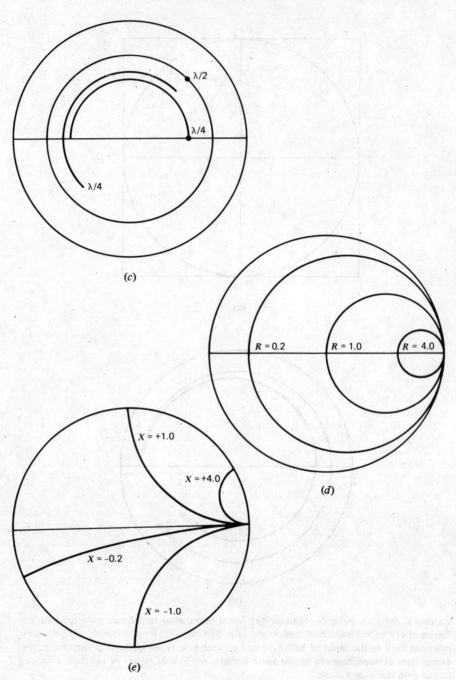

(c)

(d)

(e)

Figure 16 (*Cont.*)

termination. Therefore

$$Z_{in} = Z(-h) = \frac{1 + \Gamma_L e^{-2\gamma h}}{1 - \Gamma_L e^{-2\gamma h}} \tag{3.39}$$

Solving (3.39) for Γ_L yields

$$\Gamma_L = \left(\frac{Z_{in} - 1}{Z_{in} + 1}\right) e^{-2\gamma h} = \Gamma_s e^{-2\gamma h} \tag{3.40}$$

or,

$$\Gamma_s = \Gamma_L e^{2\gamma h} = \Gamma_L e^{2\alpha h} e^{j4\pi h/\lambda} \tag{3.41}$$

Equation 3.41 relates the reflection coefficient at one end of a line to the reflection coefficient at the other end of the line, in wavelengths. Returning to the plot of Γ, (3.41) represents a rotation and a change of radius. For a lossless line, $\alpha = 0$ and Γ_s can be found by locating Γ_L and rotating in the clockwise direction $720h/\lambda°$, as shown in Figure 16b. Conversely, from (3.40),

$$\Gamma_L = \Gamma_s e^{-2\alpha h} e^{-j4\pi h/\lambda} \tag{3.42}$$

For a lossless line, Γ_L is found from Γ_s by rotating in a counterclockwise direction, $720h/\lambda°$. These clockwise and counterclockwise rotations are usually referred to as "toward the generator" and "toward the load," respectively.

In the case of a lossy line, the rotation spirals inward (or outward) as determined by the exponential term in α.

Several important properties of lossless (or low loss) lines in the sinusoidal steady state are easily shown in terms of the rotations above: a reflection coefficient (therefore a terminating impedance) is seen identically when a lossless line is an integral multiple number of half-wavelengths long (Figure 16c). A reflection coefficient repeats in magnitude but reverses in sign every quarter-wavelength (Figure 16c). This means that a quarter-wavelength line that is shorted at the far end will look open, and vice versa.

The most general use of the Smith chart, as this mapping is known, is obtained by overlaying curves of constant R and X, where $Z_L = R + jX$. This is done by solving (3.37) for Z_L,

$$Z_L = R + jX = \frac{1 + \Gamma_L}{1 - \Gamma_L} \tag{3.43}$$

and separating the real and imaginary parts:

$$R = \frac{1 - (u^2 + v^2)}{(1 - u^2) + v^2} \tag{3.44}$$

$$X = \frac{2v}{(1 - u^2) + v^2} \tag{3.45}$$

where u and v are the real and imaginary parts, respectively, of Γ_L.

Since the two equations above are quadratic in u and v, curves of constant R and X must be conic sections. Rewriting these equations in standard form for conic sections reveals that these curves are circles:

$$\left(u - \frac{R}{1+R}\right)^2 + v^2 = \frac{1}{(1+R)^2} \tag{3.46}$$

$$(u-1)^2 + \left(v - \frac{1}{X}\right)^2 = \frac{1}{X^2} \tag{3.47}$$

Consider first several circles of constant R (Figure 16d). These circles have centers at $[R/(1+R), 0]$, and radii of $1/(1+R)$. The centers of the circles lie on the u axis, and the circles intercept the u axis at

$$u = \frac{R}{1+R} \pm \frac{1}{1+R} = \frac{R-1}{R+1} \quad \text{and } 1$$

This means that for every value of R greater than zero, circles of constant R will lie within the unit circle and will all meet at $(u, v) = (1, 0)$.

Circles of constant X, on the other hand, do not all lie within the unit circle. They have centers at $[1, 1/X]$, and radii $1/X$. Since only the sections that pass through the unit Γ circle are of interest, the sections outside the unit circle are not considered. Figure 16e shows several (sections of) circles of constant X. Note that all these circles pass through the point $(1, 0)$, as did the circles of constant R. Also, these circles all lie above the u axis for positive X and below the u axis for negative X; $X = 0$ lies on the u axis.

Figure 17 is the composite drawing of the unit circle, several circles of constant R, and several circles of constant X. The circumference is labeled in degrees away from $(1, 0)$, which denote the angle of Γ, and also fractions of a wavelength toward the generator and the load, as explained above. This composite drawing is known as an impedance Smith chart. It is possible to construct an analogous chart using admittance variables (an admittance Smith chart).

The uses of the Smith chart are too numerous to detail here. The references cited at the end of this chapter contain a myriad of Smith chart examples and techniques. Two of the more common uses are described below It should be noted that the Smith chart relationships were derived assuming a transmission line with $Z_0 = 1$. This is a generalization rather than a special case, since any system can be normalized to $Z_0 = 1$. Two applications of the Smith chart are as follows:

1. The impedance $R + jX$ is located on the Smith chart at the intersection of the correct constant-R and constant-X circles. The location of this point in the Γ plane immediately gives Γ and VSWR.

2. The input impedance of a length of line terminated in some Z_L is found by

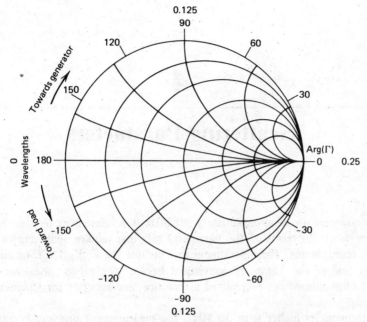

Figure 17 The composite impedance Smith chart.

first locating Z_L on the Smith chart, as described above, then rotating the point found about the center of the unit circle the correct number of wavelengths toward the generator. If the line is lossy, the rotation will also spiral inward, with the new radius given in terms of the original radius by (3.42), as follows:

$$r \equiv |\Gamma_L| = |\Gamma_s|e^{-2\alpha h} \tag{3.48}$$

3.4 SUGGESTED FURTHER READING

1. M. S. Ghausi, *Principles and Design of Linear Active Circuits*, McGraw-Hill, New York, 1965. The various two-port parameter sets and their properties are treated rigorously, as are the definitions of the different gain relations that are commonly used.

2. R. W. P. King, *Transmission Line Theory*, McGraw-Hill, New York, 1955. A very detailed and thorough treatment of impedance matching using the Smith chart, along with several examples of the less common rectangular impedance chart.

3. S. Ramo et al., *Field and Waves in Communications Electronics*, Wiley, New York, 1965. Contains a very lucid treatment of the material of this chapter, along with some useful tables.

4

Scattering Parameters

At frequencies up to approximately 30 MHz it is convenient to describe two-port networks in terms of the parameter sets that require open- and/or short-circuit terminations. These parameter sets include the Z, Y, $ABCD$ (chain), and h, sets, and so on. These are convenient because the proper open- and short-circuit terminations can be realized in practice, and accurate measurements can be made.

At frequencies higher than 30 MHz, the measurement problem becomes increasingly difficult: a physical short-circuit termination (e.g., at one end of a transmission line) requires a finite length conductor and usually exhibits inductance. A physical open-circuit termination usually exhibits capacitance due to electric field fringing. It is difficult to measure current without disturbing the circuit being measured. The "wires" connecting the measuring equipment to the system being measured have network parameters of their own, and also couple into the system being measured in many unpredictable ways.

There is a clear need for a two-port parameter set that is defined in terms of measurements that can be performed accurately at high rf and microwave frequencies. The scattering parameters answer this need. Scattering parameters are defined in terms of incident and reflected waves, and they require transmission lines that are terminated in their characteristic impedance as the boundary (terminal) conditions.

One of the advantages of the scattering matrix approach is that characteristic impedance terminations can be placed at an arbitrary distance from a network on the end of a transmission line. The signal measured at the termination will show a phase dependence on the length of the connecting line, whereas the amplitude level will be fairly independent of the length of the connecting line. This phase dependence (and any small amplitude dependence) can be "canceled" from measurements by making measurements that are not absolute, but comparative. The comparison, or reference, signal is passed along a line as nearly

identical as possible to the connecting line discussed above, but with no unknown network inserted in its path. The differences in amplitude and phase
signal levels are now the measured parameters. This approach to network analysis
is implemented in the network analyzer measurement system, which is based
entirely on making comparative scattering parameter measurements.

Another approach to microwave measurements is to automate the slotted line
technique—that is, to build multiport sampling structures that measure power
flow along a line at various fixed positions. A careful calibration of the sampling
structure yields a set of coefficients that allow a small computer to calculate
quickly reflection and transmission coefficients at a port of the structure. This
network analysis philosophy does not depend on precision microwave components, therefore finds favor at higher microwave frequencies where precision
broadband components are extremely difficult to realize. The tradeoff here is
the investment in computing capability and careful calibration procedures.

4.1 THE SCATTERING MATRIX

Consider again (3.1) and (3.2):

$$V(z) = A_1 e^{-\gamma z} + A_2 e^{+\gamma z} \tag{3.1}$$

$$I(z) = \frac{A_1 e^{-\gamma z} - A_2 e^{+\gamma z}}{Z_0} \tag{3.2}$$

Assume that at $z = 0$ a transmission line is driven by a voltage source with internal impedance Z_0 and is terminated at some arbitrary point $z \neq 0$ with an
arbitrary termination Z_L. At the point z, A_1 and A_2 represent the incident and
reflected voltages, respectively. Renaming A_1 and A_2 as V_i and V_r, and inverting
the equations above, we have

$$V_i e^{-\gamma z} = \tfrac{1}{2} [V(z) + Z_0 I(z)] \tag{4.1}$$

$$V_r e^{+\gamma z} = \tfrac{1}{2} [V(z) - Z_0 I(z)] \tag{4.2}$$

Let us divide both sides of these equations by $\sqrt{Z_0}$, and define the incident
and reflection parameters a and b by the relations

$$a \equiv \frac{V_i e^{-\gamma z}}{\sqrt{Z_0}} = \tfrac{1}{2} \left[\frac{V(z)}{\sqrt{Z_0}} + \sqrt{Z_0}\, I(z) \right] \tag{4.3}$$

$$b \equiv \frac{V_r e^{\gamma z}}{\sqrt{Z_0}} = \tfrac{1}{2} \left[\frac{V(z)}{\sqrt{Z_0}} - \sqrt{Z_0}\, I(z) \right] \tag{4.4}$$

The ratio b/a, from the above, is

$$\frac{b}{a} = \frac{V_r e^{\gamma z}}{V_i e^{-\gamma z}} = \frac{V_r}{V_i} e^{-2\gamma z} \tag{4.5}$$

The numerator in (4.5) is the reflected voltage at the termination. The denominator is the incident voltage at the termination. The ratio b/a is therefore Γ, the reflection coefficient (at the termination). Note that this result would have been the same regardless of whether (4.1) and (4.2) had or had not been divided by $\sqrt{Z_0}$, or any other number. The term $\sqrt{Z_0}$ is referred to as a normalizing factor. It is arbitrary and in this case has been chosen to put a and b into units of (power)$^{1/2}$. The usefulness of this choice will become apparent shortly.

Solving (4.3) and (4.4) in terms of V and I yields

$$V = (a + b)\sqrt{Z_0} \tag{4.6}$$

$$I = \frac{a - b}{\sqrt{Z_0}} \tag{4.7}$$

The power that is dissipated in the termination is given by

$$P = \tfrac{1}{2} VI^* = \tfrac{1}{2}(aa^* - bb^*) = \tfrac{1}{2}(|a|^2 - |b|^2) \tag{4.8}$$

The terms in (4.8) show that the power dissipated in the termination is equal to the incident power $\tfrac{1}{2}aa^*$, minus the reflected power $\tfrac{1}{2}bb^*$. Obviously the conservation of energy demonstrated here is not dependent on the correct choice of a normalizing factor. The simple form of (4.8), on the other hand, is in part due to the judicious choice of normalizing factor.

The parameters of the scattering matrix are derived by extending the concepts of incident and reflection parameters to a two-port network. In general, it is not necessary to consider both ports of the network to be connected to transmission lines of the same characteristic impedance. The discussion below, however, is limited to this case. The references at the end of the chapter include discussions of the more general situation.

Define a_1 and a_2 as the incident parameters and b_1 and b_2 as the reflection parameters at ports 1 and 2, respectively, of a linear two-port network. Following conventional two-port network current and voltage conventions (Figure 18),

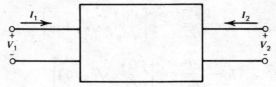

Figure 18   Two-port parameter set conventions for a length of transmission line.

(4.3) and (4.4) extend to

$$a_1 = \tfrac{1}{2}\left(\frac{V_1}{\sqrt{Z_0}} + \sqrt{Z_0}\, I_1\right) \qquad (4.9)$$

$$a_2 = \tfrac{1}{2}\left(\frac{V_2}{\sqrt{Z_0}} + \sqrt{Z_0}\, I_2\right) \qquad (4.10)$$

$$b_1 = \tfrac{1}{2}\left(\frac{V_1}{\sqrt{Z_0}} - \sqrt{Z_0}\, I_1\right) \qquad (4.11)$$

$$b_2 = \tfrac{1}{2}\left(\frac{V_2}{\sqrt{Z_0}} - \sqrt{Z_0}\, I_2\right) \qquad (4.12)$$

The scattering (S) parameters are defined by the equations

$$b_1 = S_{11}a_1 + S_{12}a_2 \qquad (4.13)$$

$$b_2 = S_{21}a_1 + S_{22}a_2 \qquad (4.14)$$

Or, in matrix notation,

$$\begin{bmatrix} S_{11} & S_{12} \\ S_{21} & S_{22} \end{bmatrix}\begin{bmatrix} a_1 \\ a_2 \end{bmatrix} = \begin{bmatrix} b_1 \\ b_2 \end{bmatrix} \qquad (4.15)$$

The four S parameters are evaluated in terms of the defining equations. Taking these in order, we begin with

$$S_{11} = \frac{b_1}{a_1}\bigg|_{a_2=0} \qquad (4.16)$$

The parameter S_{11} is determined by applying an incident wave a_1 to port 1 and measuring b_1, the amplitude of the wave emanating from port 1, with the boundary condition such that there is no incident wave at port 2. Referring to the discussion of reflection coefficients, if port 2 is terminated in Z_0, any power emanating from port 2 will be absorbed in the termination. Therefore no power will be reflected by the termination back into port 2, and the boundary condition is satisfied. In other words, S_{11} is simply the reflection coefficient measured at port 1 when port 2 is terminated in Z_0.

Similarly,

$$S_{22} = \frac{b_2}{a_2}\bigg|_{a_1=0} \qquad (4.17)$$

where S_{22} is the reflection coefficient measured at port 2 when port 1 is terminated in Z_0.

Each of the parameters S_{11} and S_{22} measures a reflection at one port of the network; S_{12} and S_{21}, on the other hand, measure transfer of power through the network:

$$S_{21} = \frac{b_2}{a_1}\bigg|_{a_2=0} \qquad (4.18)$$

where S_{21} is the ratio of the amplitude of the wave emanating from port 2 to the amplitude of an incident wave at port 1 under the condition that port 2 is terminated in Z_0. As was the case in the determination of S_{11}, the Z_0 termination at port 2 assures that $a_2 = 0$. From the definition of a_2 in (4.10), if $a_2 = 0$, then

$$V_2 = -Z_0 I_2 \qquad (4.19)$$

and using (4.12),

$$b_2 = \frac{V_2}{\sqrt{Z_0}} \qquad (4.20)$$

If the driving source is a Thévénin equivalent voltage source consisting of an ideal voltage source V_{g1} in series with a source resistance Z_0, then $V_1 = V_{g1} - Z_0 I_1$, and using (4.20) and (4.9), we write

$$a_1 = \frac{1}{2}\left[\frac{V_{g1} - Z_0 I_1}{\sqrt{Z_0}} + \sqrt{Z_0}\, I_1\right] = \frac{V_{g1}}{2\sqrt{Z_0}} \qquad (4.21)$$

Consequently

$$S_{21} = \frac{2V_2}{V_{g1}} \qquad (4.22)$$

and using the same reasoning,

$$S_{12} = \frac{2V_1}{V_{g2}} \qquad (4.23)$$

Now S_{21} and S_{12} are seen to be voltage transfer ratios. For a reciprocal network, $S_{12} = S_{21}$.

In one important respect the S parameters differ conceptually from the conventional lumped element sets such as the Z and Y parameters. The difference is that there is a characteristic impedance Z_0, assumed implicitly in the definitions of the S parameters. The value of Z_0 will appear only in analyses that convert from one parameter set to another, but the values of the S parameters for a given two-port network will be strongly dependent on the choice of Z_0 in the measuring system.

As an example of the preceding definitions and comment, consider the S parameters for a transmission line of characteristic impedance Z_0, propagation constant γ, and length h. Clearly there are no reflections at either port (either end of the line, in this case), and $S_{11} = S_{22} = 0$. Parameter $S_{21} = 2V_2/V_{g1}$ with the boundary (port 2 termination) satisfied. In this case, $V_{g1} = 2V_1$, and $V_2 = V_1 e^{-\gamma h}$. Therefore $S_{21} (= S_{12}) = e^{-\gamma h}$. The scattering parameter matrix for this length of line is

$$S = \begin{bmatrix} 0 & e^{-\gamma h} \\ e^{-\gamma h} & 0 \end{bmatrix} \tag{4.24}$$

As a second example, consider a length of transmission line with a characteristic impedance Z_a being measured with a Z_0 reference system. In this case $S_{11} = S_{22}$ (the line is symmetric), but in general is not zero. To calculate S_{11} it is necessary to find the input impedance seen by the (Thévenin) source in terms of the line parameters of characteristic impedance, propagation constant and length, and the reference (measuring) impedance Z_0. Then the reflection coefficient at port 1 can be calculated directly. To calculate S_{21} (= S_{12}), it is necessary to find V_2 and I_1 in terms of V_{g1}. These in turn are found in terms of the line parameters and Z_0. Leaving the details of these calculations as an exercise, we simply write

$$S_{11} = S_{22} = \frac{(Z_a^2 - Z_0^2)\sinh(\gamma h)}{2Z_a Z_0 \cosh(\gamma h) + (Z_a^2 + Z_0^2)\sinh(\gamma h)} \tag{4.25}$$

$$S_{12} = S_{21} = \frac{2Z_a Z_0}{2Z_a Z_0 \cosh(\gamma h) + (Z_a^2 + Z_0^2)\sinh(\gamma h)} \tag{4.26}$$

As can be seen, scattering parameters can be very simple when transmission lines and/or transmission line networks are studied in a system that has the same characteristic impedance as the network being investigated, but very complicated and unwieldy in the general case. Also, an examination of how the four S parameters in (4.25) and (4.26) change with changing Z_a will show very poor incremental sensitivity when Z_a is very different from Z_0. In terms of practical measurements, this means that S-parameter measurements are typically accurate only when the parameters being measured are not very different from Z_0.

4.2 BASIC PROPERTIES OF S PARAMETERS

Equation 4.8 showed that the power delivered to port 1 is $P_1 = \frac{1}{2}(a_1 a_1^* - b_1 b_1^*)$. Similarly, the power delivered to port 2 is $P_2 = \frac{1}{2}(a_2 a_2^* - b_2 b_2^*)$. The total power delivered to the two-port network, which for a passive network must not be less than zero, is

$$P_{in} = \frac{1}{2}(a_1 a_1^* - b_1 b_1^* + a_2 a_2^* - b_2 b_2^*) \geqslant 0 \tag{4.27}$$

In matrix notation this can be written

$$P_{in} = \tfrac{1}{2} (a^{*T}a - b^{*T}b) \geqslant 0 \qquad (4.28)$$

where a = a column matrix (vector) with elements a_1, a_2, etc.
$\quad\quad\ b$ = a column matrix (vector) with elements b_1, b_2, etc.
$\quad\quad\ T$ = matrix transpose operation

Since $b = Sa$, by basic matrix operations, $(b^*)^T = (a^*)^T(S^*)^T$. Therefore

$$2P_{in} = a^{*T}a - a^{*T}S^{*T}Sa = a^{*T}(I - S^{*T}S)\,a \geqslant 0 \qquad (4.29)$$

where I is the identity matrix.

Equation 4.29 will hold only if

$$\det\,[I - S^{*T}S] \geqslant 0 \qquad (4.30)$$

In the special case of a lossless network, the equality sign holds in (4.30), and

$$S^{*T}S = I \qquad (4.31)$$

A matrix satisfying the relation above is called a unitary matrix. Examining (4.31) term by term, we have

$$S_{11}^* S_{11} + S_{21}^* S_{21} = 1 \qquad (4.32a)$$

$$S_{12}^* S_{11} + S_{22}^* S_{21} = 0 \qquad (4.32b)$$

$$S_{11}^* S_{12} + S_{21}^* S_{22} = 0 \qquad (4.32c)$$

$$S_{12}^* S_{12} + S_{22}^* S_{22} = 1 \qquad (4.32d)$$

Note that (4.32c) and (4.32b) are complex conjugates.

For a reciprocal network, $S_{21} = S_{12}$, and

$$|S_{11}(j\omega)|^2 + |S_{21}(j\omega)|^2 = 1 \qquad (4.33a)$$

$$|S_{22}(j\omega)|^2 + |S_{21}(j\omega)|^2 = 1 \qquad (4.33b)$$

From the two equations above it follows that for a reciprocal network,

$$|S_{11}(j\omega)| \leqslant 1 \qquad (4.34)$$

$$|S_{22}(j\omega)| \leqslant 1 \qquad (4.35)$$

$$|S_{21}(j\omega)| \leqslant 1 \qquad (4.36)$$

Also, for a lossless network, if $S_{21} = 0$ no power is transmitted through the network from port 1 to port 2 (the forward direction); therefore $|S_{11}| = 1$. Similarly, for a lossless network, when $S_{12} = 0$, $|S_{22}| = 1$.

An interesting property of S parameters is that although there exist networks that do not have a set of Z parameters, and networks that do not have a set of Y parameters (meaning that some of the terms are not finite), it has been proved that all passive networks have S parameters.

It has been shown that S_{11} and S_{22} are the reflection coefficients at ports 1 and 2, respectively, when the opposite port is properly terminated. Parameters S_{21} and S_{12} are related to the forward and reverse insertion losses, defined in the following manner.

Consider a source V_g with source impedance Z_0, driving a load impedance Z_0. Clearly the voltage across the load is $V_g/2$. Consider the case of the same source and load, but with a two-port network inserted between them. Using (4.23), the voltage across the load is now $V_g S_{21}/2$. Define the insertion voltage ratio as the ratio of the voltage across the load with the two-port network in place to the voltage across the load without the two-port network in place:

$$\text{insertion voltage ratio} = \frac{V_g S_{21}/2}{V_g/2} = S_{21} \qquad (4.37)$$

Since the power dissipated in the load is proportional to the square of the magnitude of the voltage across the load, it is possible to define the insertion power ratio as $|S_{21}|^2$. Often the reciprocal of the insertion power ratio, in decibels, is defined as the insertion loss:

$$\text{insertion loss} = -10\ \text{Log}_{10}\ |S_{21}|^2 = -20\ \text{Log}_{10}\ |S_{21}| \qquad (4.38)$$

4.3 EXAMPLES OF TWO–PORT NETWORKS CHARACTERIZED BY S PARAMETERS

Example 1. The matched attenuator (pad). A matched attenuator, or "pad," is a lossy passive device that presents an input impedance Z_0 when it is terminated in Z_0, and vice versa. If the pad is physically small enough that the phase shift through it is negligible at frequencies of interest, then

$$S = \begin{bmatrix} 0 & k \\ k & 0 \end{bmatrix} \qquad (4.39)$$

The coefficient k is a measure of the insertion loss of the attenuator. From (4.38) we have

$$\text{insertion loss} = -20\ \text{Log}\ (k) \qquad (4.40)$$

Example 2. The ideal isolator. An isolator is a nonreciprocal device that allows power flow in only one direction. An ideal isolator has no internal losses in the allowed or "forward" direction. Again, ignoring internal phase shifts,

$$S = \begin{bmatrix} 0 & 0 \\ 1 & 0 \end{bmatrix} \qquad (4.41)$$

This matrix, although very simple in appearance, describes a fairly involved device concept. Expanding (4.41) and writing out the full equations, we have

$$b_1 = S_{11}a_1 + S_{12}a_2 = 0 \tag{4.42a}$$

$$b_2 = S_{21}a_1 + S_{22}a_2 = a_1 \tag{4.42b}$$

Equation 4.42a shows that no impinging wave or combination of waves on either port of the isolator will cause a nonzero b_1 term. Port 1 appears as a simple Z_0 termination under all conditions. Equation 4.42b says that $b_2 = a_1$. In other words, the isolator is transparent to a wave impinging on port 1.

To calculate the power dissipated in the ideal isolator, we use

$$P_{\text{diss}} = \tfrac{1}{2} [a_1 a_1^* - b_1 b_1^* + a_2 a_2^* - b_2 b_2^*] = \tfrac{1}{2} a_2 a_2^* \tag{4.43}$$

Equation 4.43 shows that any incident signal at port 2 is dissipated entirely within the isolator. Also, no part of a signal incident at port 1 is dissipated within the isolator—unless it is reflected back into port 2 by a mismatch at that port. Insertion losses and mismatch losses in a network containing an isolator (or any nonreciprocal device) must be calculated carefully; since intuition and experience based on reciprocal networks can be very misleading.

The theory and notation of S parameters extends quite readily to multiport networks. The relationships developed in Section 4.2 also extend readily. For an N port network, S_{ii} is the reflection coefficient at port i when all the other ports are properly terminated. Then S_{ij}, $i \neq j$, is twice the voltage at port i due to a source at port j when all ports except j are terminated in Z_0, and port j is driven by a source with internal impedance Z_0. As in the case of a two-port network, for a reciprocal network, $S_{ij} = S_{ji}$.

Example 3. The three-port circulator. A circulator is a nonreciprocal device having three or more ports; it transmits the power incident at any given port to the next port in rotational order. A circulator must therefore have a clockwise or counterclockwise "sense," which is usually marked on the circulator. For a lossless three-port circulator, ignoring internal phase shifts,

$$S = \begin{bmatrix} 0 & 0 & 1 \\ 1 & 0 & 0 \\ 0 & 1 & 0 \end{bmatrix} \tag{4.44}$$

If one port of a three-port circulator is terminated in Z_0, the remaining two-port network has the scattering matrix given in (4.41) and is, in fact, an ideal isolator.

In practice many isolators are built by terminating one port of a three-port circulator. In this case it is clear where the power dissipation inside the isolator occurs.

Example 4. The bidirectional coupler. When the fields surrounding the conductors of two (or more) transmission lines interact, there must be some coupling between these lines. Chapter 8 gives a detailed description of the physical properties of coupled lines. The significant property of coupled transmission lines (or modes of propagation in general) that distinguishes coupled lines from coupled resonators (modes of oscillation) is the existence of preferred directions of power flow in the coupling mechanisms.

The lossless, reciprocal, bidirectional coupler is, in its simplest form, a pair of unshielded transmission lines held in close proximity for some distance.

Ignoring phase shifts, the scattering matrix formed by the two lines connected as shown in Figure 19 is approximately

$$S = \begin{bmatrix} 0 & 1 & k & 0 \\ 1 & 0 & 0 & k \\ k & 0 & 0 & 1 \\ 0 & k & 1 & 0 \end{bmatrix} \tag{4.45}$$

where k is a measure of the line-to-line coupling.

Usually, as with the attenuator, k is expressed in decibels. For example, when speaking of a "30 dB coupler, $-20 \, \text{Log}\,(k) = 30$. This means that when a signal travels from port 1 to port 2, a sample of that signal that has been attenuated by 30 dB will appear at port 3, but not at port 4.

Equation 4.45 is never exactly correct because of the following simplifications:

1. If some of the power traveling from port 1 to port 2 is "sampled" and dissipated in a load at port 3, then S_{21} cannot have a magnitude of 1 (or the device would be showing a power gain). For example, in a 20 dB coupler, $k = 0.1$, and $|S_{21}|$ is then ≈ 0.995. As can be seen, (4.45) is reasonable even though not exact.
2. Since the coupling mechanism is based on transmission line lengths in close proximity, the picture of a 0 degree phase shift transfer characteristic is incorrect in principle. This is not the same as the zero phase shift approximation for the

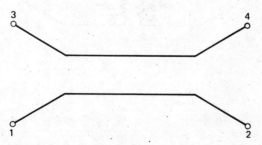

Figure 19 Schematic of a simple four-port directional coupler. Ground returns not shown.

attenuator, isolator, or circulator, since it is possible to conceive of one of these devices being built small enough that no phase shift could ever be measured. On the other hand, in the case of the coupler, the physical length is necessary for the basic coupling mechanism, so that shrinking the device to reduce phase shift is an invalid concept.

It should be mentioned in passing that all the foregoing examples were idealized in ways other than the coupling alone. For example, an isolator never has infinite reverse direction insertion loss; a directional coupler never has zero reverse direction coupling, and so on.

The accuracy of the bidirectional coupler just described as a measurement tool is limited in the following manner. Assume that some transmission line is severed and ports 1 and 2 of the bidirectional coupler of Figure 19 are inserted. Power meters connected to ports 3 and 4 of the coupler indicate the power flowing in both directions in the transmission line. However, suppose that the meters themselves do not present an exact Z_0 termination to ports 3 and 4. In this case the meters will reflect power back into the coupler. A portion of the signal incident on the meter connected to port 3 will appear at port 4, and vice versa. To minimize this problem, dual directional couplers may be constructed (Figure 20). Internal terminations, carefully matched to the lines and to each other are supplied as part of the coupler. They are not accessible from the outside, and their parameters are part of the permanent calibration of the coupler—this calibration is usually printed on the case of the coupler. This coupler has the scattering matrix

$$S = \begin{bmatrix} 0 & 1 & k & 0 \\ 1 & 0 & 0 & k \\ k & 0 & 0 & 0 \\ 0 & k & 0 & 0 \end{bmatrix} \tag{4.45}$$

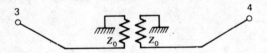

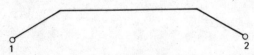

Figure 20 Schematic of an improved directional coupler.

The advantage of this coupler over the simpler scheme is that ports 3 and 4 are isolated from each other, and the unwanted "back" response cannot be aided by poor or nonrepeatable terminations at these ports. Note of course that there is no signal path from port 3 to port 4, and the signal path (ports 1 and 2) must be labeled at the terminations of the coupler.

Example 5. The three-port junction. Suppose that a three-port network was made up of three transmissions lines meeting at a common node. The scattering matrix would be

$$S = \tfrac{1}{3} \begin{bmatrix} -1 & 2 & 2 \\ 2 & -1 & 2 \\ 2 & 2 & -1 \end{bmatrix} \qquad (4.46)$$

4.4 THE NETWORK ANALYZER MEASUREMENT SYSTEM

Although absolute measurements of signal magnitudes and phases at VHF and at higher frequencies are very difficult to make, comparative measurements are relatively simple. This is because despite the near impossibility of building, for example, a directional coupler with perfectly flat response characteristics over any reasonable frequency range, it is not particularly difficult to build a closely matched pair of such units. In practice, it is not uncommon to build many units, then to sort and match pairs. The net result is the same regardless of how it came about. A system can be built of two networks that are identical except for the unknown two-port network under investigation. Identical (matched) converters heterodyne the two signals down to a convenient intermediate frequency (IF) where absolute measurements are more readily made. At the intermediate frequency a matched pair of receivers processes the two signal channels, which are referred to as the reference and test channels. An automatic gain control (AGC) signal developed in the reference channel can be used to control the gain of both channels, and the actual level of the signal source becomes unimportant. Furthermore, phase lock techniques in the frequency conversion circuits allow the system to operate under swept frequency conditions.

The test system just described is known as a microwave network analyzer. The concepts are straightforward, but a test system that is to provide accurate data over a multigigahertz range must be designed, built, and calibrated as carefully and precisely as possible. A good network analyzer is therefore a rather expensive piece of equipment.

A network analyzer system can be divided (conceptually) into two discrete systems: a dual channel receiver that presents magnitude and phase information by comparing the signals on the two channels, and an rf "test set" that provides proper terminal conditions and signal sampling outputs for examining some set of two-port parameters of an arbitrary network. There is no a priori reason for designing a network analyzer to measure scattering parameters, but as noted earlier, scattering parameters are the most accurately measurable parameters at VHF and at higher frequencies.

For the purpose of examining techniques for scattering parameter measurements, the internal workings of the dual channel receiver are of no consequence. Let us, therefore, assume the circuitry of the receiver as given, and confine our attention to the rf circuitry.

The design goals for the rf test set described above are as follows:

1. The test set must be able to establish the proper terminal conditions at an arbitrary network to measure each of the four S parameters.

2. All four configurations must have the same electrical lengths in all paths. Furthermore, this length must be adjustable to match the length of the signal path in the reference channel. This is necessary for swept frequency measurements because the phase responses of both channels, aside from the network under examination, must be identical.

3. All components must be carefully matched to their counterparts in the test and reference channels. Needless to say, these components also should be as closely matched to Z_0 as possible.

Figure 21 is a schematic diagram of a possible rf test set. The four SPDT switches shown may be coaxial switches or relays, or possibly even PIN diode switches. The dashed lines refer to adjustable length transmission lines. All resistors are actually precision Z_0 terminations. The relay coils (or diode driving circuits) are connected so that not all combinations are possible. Table 2 lists the four possible switching combinations and the scattering parameter terminal conditions that are established by each of them.

Consider the cases tabulated, ignoring for the moment the terms in parentheses in Table 2. The subscript A refers to port A being used as the input or "left-hand" port of the network being examined.

In the first case tabulated, the signal enters the test set through a small pad. This pad helps to ensure that the signal source appears to have an internal impedance Z_0, and also that the source itself does not experience drastically varying loads. The signal is routed through the upper directional coupler and into port A of the unknown network. This is path *a-b-c-d-e* in Figure 21. Any signal reflected by the input of the unknown network is reflected back along the same path and absorbed in the Z_0 termination of the pad and source. The signal incident at port A is sampled by the directional coupler and appears at the reference channel output port of the test set. This signal has traveled path *a-b-c-g-i* (Figure 21). The reflected signal is also sampled by the directional coupler, and

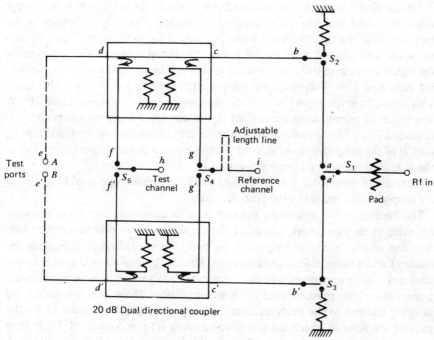

Figure 21 Simplified schematic of an *S*-parameter test set.

it appears at the test channel output port of the test set. This signal has traveled path *a-b-c-d-e-d-f-h*. If the adjustable length line in the reference channel is set so that path length *g-i* is the same as path length $2(d\text{-}e) + (f\text{-}h)$, the phase difference between the signals at the test and reference ports is identically the phase angle of S_{11} of the unknown network. Also, the ratio of the voltage levels at the test and reference channel ports is the magnitude of S_{11} of the unknown network.

Table 2 Switching Combinations for the
S-Parameter Test Set

Switch					
S_1	S_2	S_3	S_4	S_5	Measurement
1	2	2	1	1	$S_{11A}(S_{22B})$
1	2	2	1	2	$S_{21A}(S_{12B})$
2	1	1	2	2	$S_{22A}(S_{11B})$
2	1	1	2	1	$S_{12A}(S_{21B})$

Assume that the lower half of the circuit of Figure 21 is a mirror image of the upper half, and consider the second case tabulated. The only difference in the test circuits of the first two cases is the position of S_5. In the second case S_5 connects the test channel port to the lower directional coupler, where it samples the signal arriving at the coupler from port B of the unknown network, following path e'-d'-f'-h. Note that the path lengths e-d-f and e'-d'-f' are the same. This means that the adjustment of the line length in the reference channel that was made to permit measurement of S_{11} is correct for the measurement just described—S_{21}. This measurement, as can be seen, is comparing the signal leaving port B of the unknown network (test channel) to the signal incident at port A of the unknown network (reference channel).

Cases 3 and 4 in Table 2 set the conditions for measuring S_{12} and S_{22}. These are simply mirror images of the first two cases.

The "setting up" of a network analyzer for a measurement consists of properly adjusting three parameters. These are the amplitude balance between the channels, the relative electrical lengths of the two channels (phase tracking with frequency), and a phase reference calibration. Since in general an unknown network does not conform mechanically to the network analyzer ports as supplied, some transmission lines must be used for interconnection. These lines are considered as being internal to the analyzer. That is, test ports A and B in Figure 21 are the ends of the lines to which the unknown network is to be connected. These lines may be lossy and/or dispersive. They should, however, be well matched to each others. Also, calibration of amplitude and phase over a wide frequency range is simpler if these lines are relatively loss free. Ideally, the adjustable line in the reference channel should be of the same material as the test port lines. This is often not convenient, unfortunately.

Consider the following procedure for adjusting the three parameters just enumerated. In the jargon of microwave circuitry, the intent of the procedure is to establish a "reference plane" at the test port (A and B) terminals.

1. *Amplitude balance.* Either connect test ports A and B together and adjust the receiver gains for the same signal level in the test and reference channels, while the switches (relays) are set for measuring a transfer parameter (S_{12} or S_{21}), or connect good electrical short or open circuits to both ports A and B and adjust the receiver channel gains for the same signal level in both channels while set up to measure a reflection parameter (S_{11} or S_{22}). In principle, these alternative procedures are equivalent. In practice, of course, there are always some inequities in a system. Directional couplers are never ideal or exactly duplicatable. If particularly accurate measurements of any one parameter are important, the calibration should be done while set up for that particular parameter.

2. *Phase tracking.* While sweeping the signal source over the frequency range of interest, repeat either of the connection instructions for the amplitude balance

adjustment, and adjust the line length of the reference channel so that there is a constant phase difference between the two channels. The same observations concerning accuracy made in connection with amplitude balance pertain to the phase tracking adjustment. In addition, note that an open circuit created by merely leaving a coaxial connector "dangling" is a poor physical open circuit—the fringing capacitance between the center and shield terminals of the connector are in general not negligible.

3. *Phase reference*. The phase angle observed in setting the phase tracking adjustments should be calibrated to be zero degrees for a transfer or open circuit reflection calibration, or 180 degrees for a short-circuit reflection coefficient calibration.

Network analyzers are often used to measure the small signal parameters of various active devices such as transistors and PIN diodes. To measure small signal parameters it is necessary to establish some specific bias conditions in the device, then to measure the parameters while applying an rf signal whose level is not high enough to disturb the effective bias. Since by nature a network analyzer is insensitive to signal level, it is easy to vary the signal level experimentally and confirm that the small signal assumptions are not being violated.

Bias is usually introduced to the device through a pair of "bias T's", one at each test port. Figure 22 schematizes a simple bias T. A capacitor in series with the rf line serves to prevent the dc bias from "backing up" into the analyzer; an inductor in series with the bias supply serves to keep rf out of the bias supply.

The active device is usually plugged into a test fixture of some sort. This fixture must be carefully designed to look like a pair of Z_0 lines of equal length from the analyzer test ports to the device socket pins. The test fixture is then calibrated as part of the analyzer, and the fixture's socket pins become the effective test ports of the analyzer. To be able to study a three-terminal device under the three possible choices of common terminal (such as common base, emitter, or collector for a junction transistor), it is convenient to be able to choose arbitrarily which port, A or B, is the input port of the device. This distinction was not necessary when reciprocal devices were being studied. Referring to Table 2, the parameter list in parentheses shows which parameter is being

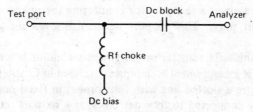

Figure 22 A simple bias T.

evaluated when port B, rather than port A, is taken to be the input port. Keep in mind that the bias voltages must be correctly connected independently of the choice of input and output rf ports.

The circuit of Figure 21 is only one of many ways to establish terminal conditions and measure S parameters. Various equivalent networks can be built using, for example, directional bridges or power splitters. It is also possible to replace the variable length reference line with a frequency-tracking phase delay in the reference channel receiver at the intermediate frequency. Computer accuracy-enhancing techniques such as storing the detailed calibration information at sampled frequency intervals and automatically using this information to correct measured data can be used to great advantage. In general, the increasing availability and decreasing cost of computing capability is causing a decrease in emphasis on precise rf connectors, couplers, and so on, and an increase in emphasis on automatic evaluation and correction of measured data based on stored knowledge of the actual rf components being used.

4.5 REFLECTOMETER NETWORK MEASUREMENTS

At microwave frequencies higher than approximately 10 GHz, it becomes difficult to build the matched pair of signal "processing" channels described above. The most difficult accuracy to maintain is that of phase (or more properly, phase difference) measurement. In an alternate approach to the network analyzer system that has been developed, one can deduce all the necessary information for a determination of a network's S parameters from signal magnitude or power level measurements alone. Consider first the problem of measuring the reflection coefficient (or equivalently, the input impedance) of a one-port network. The reflection coefficient is determined by measuring the ratio of the complex incident voltage wave to the complex reflected voltage wave at the port. The impedance could be determined directly by measuring the ratio of the complex voltage to the complex current at the port. In either case, there are two complex parameters—that is, four numbers that must be measured to characterize the one-port network.

A measurement "apparatus" capable of achieving this goal while fixed in position in terms of interconnections must have six ports. There must be an input port for a signal source, a test port for connecting the unknown network, and four measurement ports. This six-port network is given the general name of reflectometer.

Conceptually, the reflectometer measurement technique is reminiscent of the slotted line VSWR measurement techniques described in Chapter 3. One can, for that matter, picture a slotted line with four probes in fixed positions along the line, permanently connected to four detectors, as a six-port reflectometer. The

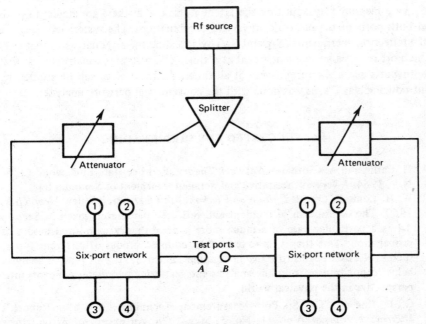

Figure 23 Dual reflectometer network analyzer.

calculations required to convert the four detector output voltage levels to the impedance being measured are reasonably involved—for hand calculation. On the other hand, once the system has been built and calibrated, the calculations are a known task that is easily accomplished by a computerized data acquisition and processing system. In practice, slotted lines are of no real use as reflectometers—at the frequencies where the measurements are accurate, better techniques are available, and at the high microwave frequencies where reflectometer techniques are needed, the slotted line is far too imprecise a tool.

For actual microwave measurements, a reflectometer is built up of various power splitters, couplers, isolators, and other components. In principle, the number of combinations of networks that are usable as reflectometers is infinite, and to date no formal optimization procedure has been devised for designing the best. The calculations required to convert the reflectometer measurements to impedance (or reflection coefficient) values are functions of the actual reflectometer design. They are usually quite involved, but again it should be emphasized that computing capability is an inexpensive commodity.

The reflectometer technique may be extended to two-port network measurements by building a system that has two reflectometers—one at each port of the two-port network being studied. Figure 23 shows a possible reflectometer network analyzer (two-port network) system. The transfer parameters (S_{12} and

S_{21}) are measured by adjusting the attenuators so that there are incident signals at both ports of the network, in varying combinations. The system will measure the reflection coefficients (S_{11} and S_{22}) by adjusting the attenuators so that only one port receives an incident signal at a time. The system is constructed so that both ports see a Z_0 termination at all times. Active devices can be studied by introducing bias T's, as was done with the conventional network analyzer.

4.6 SUGGESTED FURTHER READING

1. H. Carlin and A. Giordano, *Network Theory*, Prentice-Hall, Englewood Cliffs, N.J., 1964. A very authoritative and detailed treatment of S parameters.
2. W. H. Louisell, *Coupled Mode and Parametric Electronics*, Wiley, New York, 1960. The description of incident and reflected parameters given in Section 4.1 is a particular case of a much more general theory of normal modes of propagation. This theory is extended to coupled modes of different types, such as electron beams to transmission lines, in this text. This type of analysis helps to put transmission line analysis and scattering parameter concepts into perspective in the physical world.
3. G. F. Engen, "The Six-Port Measurement Technique, A Status Report," *Microwave Journal*, Vol. 21, No. 5, May 1978. An excellent introductory article on reflectometer concepts.

IEEE Transactions on Microwave Theory and Techniques, Vol. MTT-25, December 1977, assembled a series of articles representing the state of the art in reflectometer theory and technology:

C. A. Hoer, "A Network Analyzer Incorporating Two Six-Port Reflectometers."

G. F. Engen, "The Six-Port Reflectometer: An Alternative Network Analyzer."

G. F. Engen, "An Improved Circuit for Implementing the Six-Port Technique of Microwave Measurements."

M. P. Weidman, "A Semiautomated Six Port for Measuring Millimeter-Wave Power and Complex Reflection Coefficient."

H. Cronson and L. Susman, "A Six-Port Automatic Network Analyzer."

5

Some Non-TEM Transmission Lines

When a cross-sectional view of a transmission line shows an inhomogeneous dielectric material, the transmission line cannot support a pure TEM wave at frequencies other than dc. This is because the electric field lines cannot satisfy all the necessary boundary conditions at the dielectric interface and metallic surfaces without the inclusion of a longitudinal component of the electric field at the dielectric interface. It seems physically reasonable, however, that at low enough frequencies the transmission line would behave according to the capacitance and inductance values predicted by the dc calculations. This turns out to be the case, and the frequency range over which this approximation is useful is referred to as the quasi-static approximation range.

Treatment of the frequency-dependent properties of nonhomogeneous lines is beyond the scope of this book. A short discussion of the frequency-dependent case is presented for microstrip and slotline, for completeness. In the case of microstrip, a circuit model is shown that is very useful in picturing the variation of line properties with frequency.

The discussion in this chapter is limited almost exclusively to the microstrip transmission line. Microstrip is a very important practical transmission line, principally due to the ease with which it lends itself to the photolithographic fabrication technology predominant today.

Slotline is a non-TEM transmission line that also lends itself to photolithographic technologies. It has not achieved the general acceptance that microstrip has received, primarily because slotline characteristically has fields that extend quite a distance from the substrate, whereas microstrip fields are confined closely about the upper conductor and in the dielectric.

5.1 THE QUASI–STATIC MICROSTRIP LINE

The microstrip transmission line consists of a dielectric layer over a ground "plane" with a narrow center or upper conductor resting on the dielectric. Typical dielectric materials are various Teflon and glass-reinforced Teflon materials ($\epsilon_r \approx 2.5$) and alumina ($\epsilon_r \approx 10$), although materials with a relative dielectric constant as high as 100 are being used.

In Figure 24, a cross section of a microstrip line, the shielded, or enclosed, microstrip line is seen. It is also possible to discuss the unshielded, or open, microstrip line, although this case is not usually realized in practice. In most cases b is much larger than h, and the distinction is academic.

Because of the proximity of the ground plane (under the dielectric) to the upper conductor, and the remoteness of the cover and side walls from the upper conductor, almost all the electric field lines originate on the upper conductor and terminate on the ground plane. These lines pass through the dielectric and possibly also through the air. A typical electric field pattern appears in Figure 25.

It is important to remember that the non-TEM properties of microstrip are

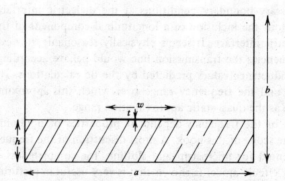

Figure 24 Cross section of an enclosed microstrip transmission line.

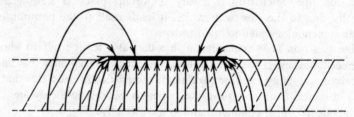

Figure 25 Typical (low frequency) electrical field pattern about a microstrip line.

caused by the existence of two different dielectric constants in the line cross section, not by the physical asymmetry of the line. If the relative dielectric constant of the dielectric layer was 1, the line shown in Figure 24 would be a TEM line. The dc capacitance in this case would be the relevant line capacitance at all frequencies, and the inductance and capacitance, L and C, would be related at all frequencies by (1.39).

For $\epsilon_r \neq 1$, C obviously must be a function of ϵ_r; however L is not. A convenient way to evaluate L when some analytic or numerical calculation procedure is available for C, is to temporarily set $\epsilon_r = 1$, find $C(\epsilon_r = 1)$, and then find L using (1.39):

$$L = \frac{\mu_0 \epsilon_0}{C(\epsilon_r = 1)} \qquad (1.39')$$

Once the correct values of L and C have been found, the characteristic impedance Z_0 is $\sqrt{L/C}$ and the wave velocity is $(LC)^{-1/2}$, as with conventional lines. Returning to (1.39) and solving for ϵ, we find that ϵ has a value lying between $\epsilon_r \epsilon_0$ and ϵ_0. This value is known as the effective dielectric constant for the line. It is a function of the dielectric material and the geometry.

The exact formulas for the characteristic impedance and effective dielectric constant for the microstrip line can be found using conformal transformations. The results are expressed in terms of elliptic integrals and are very unwieldy. The approximate formulas given below are accurate to within 1%.

For the upper conductor thickness (t) equal to 0, for $w/h \leqslant 1$,

$$Z_0 \simeq \frac{60}{\sqrt{\epsilon_{\text{eff}}}} \, \text{Ln} \left[8 \frac{h}{w} + 0.25 \frac{w}{h} \right] \qquad (5.1)$$

where

$$\epsilon_{\text{eff}} = \frac{\epsilon_r + 1}{2} + \frac{\epsilon_r - 1}{2} \left[\left(1 + \frac{12h}{w}\right)^{-1/2} + 0.04 \left(1 - \frac{w}{h}\right)^2 \right] \qquad (5.2)$$

For $w/h \gg 1$, we have

$$Z_0 \simeq \frac{120\pi/\sqrt{\epsilon_{\text{eff}}}}{w/h + 1.393 + 0.667 \, \text{Ln}(w/h + 1.444)} \qquad (5.3)$$

where

$$\epsilon_{\text{eff}} \simeq \frac{\epsilon_r + 1}{2} + \frac{\epsilon_r - 1}{2} \left(1 + \frac{12h}{w}\right)^{-1/2} \qquad (5.4)$$

Figures 26 and 27, respectively, plot Z_0 and ϵ_{eff} as a function of w/h for several values of ϵ_r. Referring to Figure 27, note that for w/h very large, $\epsilon_{\text{eff}} \approx \epsilon_r$.

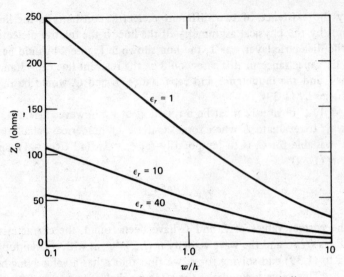

Figure 26 Z_0 versus w/h for the quasi-static microstrip line.

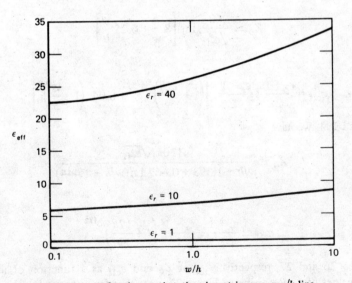

Figure 27 ϵ_{eff} for the quasi-static microstrip versus w/h line.

This is because a large w/h means that the line capacitance is essentially the parallel plate capacitance of the two conductors. On the other hand, for w/h very small Figure 26 shows that Z_0 grows slowly with decreasing w/h. This means that, in practice, there are definite limits to how high Z_0 can be made on a given thickness of a given dielectric material.

As the thickness of the upper conductor increases, electric field lines from the ground plane will reorient and terminate along the vertical edge of the upper conductor. This reorientation causes the capacitance of a given width upper conductor system to increase with increasing upper conductor thickness. The formulas above may be corrected for this effect by introducing an effective width, zero thickness, upper conductor: for $w/h \geqslant 1/2\pi$,

$$w_{\text{eff}} = w + \frac{t}{\pi} \left(1 + \text{Ln}\, \frac{2h}{t} \right) \tag{5.5}$$

and for $w/h \leqslant 1/2\pi$,

$$w_{\text{eff}} = w + \frac{t}{\pi} \left(1 + \text{Ln}\, \frac{4\pi w}{t} \right) \tag{5.6}$$

The wave velocity of a microstrip line, therefore its electrical length, must be calculated using ϵ_{eff}. This means that v is a function not only of the materials, but also of the geometry of the system. This conclusion is completely unparalleled in TEM line theory.

When a microstrip line is built on a magnetic dielectric such as a ferrite, there is a nonuniform permeability as well as a nonuniform permittivity to contend with. As would be expected, an effective permeability can be calculated in a manner analogous to the effective permittivity calculation.

5.2 THE FREQUENCY-DEPENDENT MICROSTRIP LINE

As frequency increases, the longitudinal electric field component becomes significant, and the quasi-static microstrip approximation loses its validity. A full discussion of the frequency dependence of microstrip line is beyond the scope of this treatment. The basic properties of the frequency-dependent microstrip line can be derived in terms of an experimental parameter, however, by considering the microstrip line to have a circuit model consisting of the usual TEM line circuit model coupled to a line model that cuts off below some finite frequency. Implicit here is the assumption that the quasi-static model begins to fail when it couples appreciable to *one* particular wave-guide mode, with coupling to all other modes neglected. This assumption is not justified here, but merely taken as a plausible given.

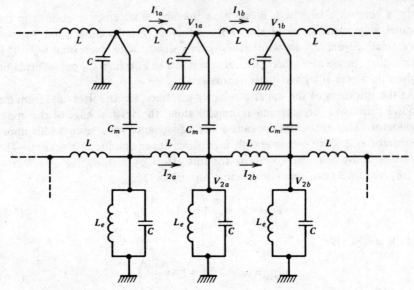

Figure 28   Circuit model for the frequency-dependent microstrip line.

Consider the circuit of Figure 28. Voltage and current variables V_1 and I_1 refer to the TEM circuit, variables V_2 and I_2 refer to the wave guide circuit. The two lines are capacitively coupled by the distributed capacitance C_m. The four circuit equations that describe figure 28 are as follows:

$$\frac{i_{1a} - i_{1b}}{\Delta z} = j\omega C V_{1a} + j\omega C_m V_{2a} \tag{5.7}$$

$$\frac{V_{1a} - V_{1b}}{\Delta z} = j\omega L i_{1b} \tag{5.8}$$

$$\frac{i_{2a} - i_{2b}}{\Delta z} = j\left(\omega C - \frac{1}{\omega L_C}\right)V_{2a} + j\omega C_m V_{1a} \tag{5.9}$$

$$\frac{V_{2a} - V_{2b}}{\Delta z} = j\omega L i_{2b} \tag{5.10}$$

Letting $\Delta z \to 0$ as was done in Chapter 1 for the basic TEM line, we have

$$\frac{\partial i_1}{\partial z} = -j\omega(C_1 V_1 + C_m V_2) \tag{5.11}$$

$$\frac{\partial V_1}{\partial z} = -j\omega L i_1 \tag{5.12}$$

$$\frac{\partial i_2}{\partial z} = -j\omega \left[\left(C - \frac{1}{\omega^2 L_e} \right) V_2 + C_m V_1 \right] \tag{5.13}$$

$$\frac{\partial v_2}{\partial z} = -j\omega L i_2 \tag{5.14}$$

Assuming that the line is infinitely long and that a wave must propagate with a uniquely defined propagation constant, γ, let

$$V_1 = e^{-\gamma z} \tag{5.15}$$

$$i_1 = \frac{e^{-\gamma z}}{Z_{01}} \tag{5.16}$$

$$V_2 = A e^{-\gamma z} \tag{5.17}$$

$$i_2 = \frac{A e^{-\gamma z}}{Z_{02}} \tag{5.18}$$

Substituting the four equations above into the four differential equations (5.11) to (5.14), we write

$$\frac{\gamma}{Z_{01}} = j\omega \left[C + A C_m \right] \tag{5.19}$$

$$\gamma = \frac{j\omega L}{Z_{01}} \tag{5.20}$$

$$\frac{A\gamma}{Z_{02}} = -j\omega \left[\left(C - \frac{1}{\omega^2 L_e} \right) A + C_m \right] \tag{5.21}$$

$$\gamma A = j\omega L \frac{A}{Z_{02}} \tag{5.22}$$

Equations 5.20 and 5.22 show that $Z_{01} = Z_{02}$. Substituting (5.20) into (5.19) and (5.21) yields

$$\gamma^2 = -\omega^2 L (C + A C_m) \tag{5.23}$$

Eliminating A between the two equations above and solving for γ^2, we have

$$\gamma^2 = \frac{k^2 - 2\omega^2 LC \pm \sqrt{k^4 + (2\omega^2 LC_m)^2}}{2} \tag{5.24}$$

where

$$k \equiv \frac{L}{L_e} \tag{5.25}$$

To gain some insight into the modes of propagation, consider the two lines in an uncoupled situation—that is, let $C_m = 0$ in (5.24):

$$v^2 \Big|_{C_m=0} = \frac{-\omega^2}{\gamma^2}\Big|_{C_m=0} = \frac{-2\omega^2}{k^2 - 2\omega^2 LC \pm k^2} = \begin{cases} \dfrac{1}{LC - k^2/\omega^2} & + \text{ case} \\[2ex] \dfrac{1}{LC} & - \text{ case} \end{cases}$$

(5.26)

Uncoupled, the − sign corresponds to the TEM line, and the + sign corresponds to a wave guide mode with a low frequency cutoff at

$$\omega_{\text{cutoff}} = \frac{k}{\sqrt{LC}}$$

(5.27)

Since the $C_m = 0$ (uncoupled) case shows that the − sign in (5.24) corresponds to the TEM line, consider this case further for a moment. Solving (5.24) for v^2 and considering only the − sign,

$$v^2 = \frac{-2\omega^2}{k^2 - 2\omega^2 LC - \sqrt{k^4 + (2\omega^2 LC_m)^2}}$$

(5.28)

At $\omega = 0$ (5.28) is indeterminate. Using L Hôpital's rule, therefore,

$$\lim_{\omega \to 0} v^2 = v_{\text{dc}}^2 = \frac{1}{LC}$$

(5.29)

The values of L and C are known for the dc (quasi-static) case, as was discussed in Section 5.1.

As ω goes to infinity,

$$v^2 \longrightarrow \frac{2\omega^2}{2\omega^2 L(C + C_m)}$$

(5.30)

or,

$$v \xrightarrow{\omega \to \infty} \frac{1}{\sqrt{L(C + C_m)}}$$

(5.31)

As $\omega \to \infty$ the fields increasingly confine themselves between the conductors— that is, in the dielectric,

$$\lim_{\omega \to \infty} (v) \longrightarrow \frac{v_0}{\sqrt{\epsilon_r}}$$

(5.32)

The only parameter in (5.28) that is not known at this point is k. Unfortu-

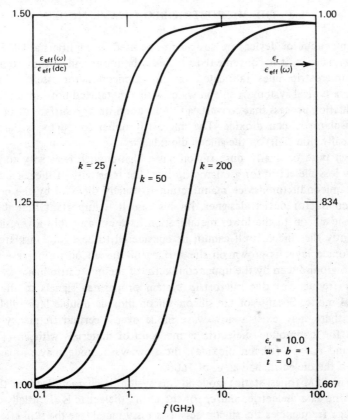

Figure 29 $\epsilon_{eff}(\omega)/\epsilon_{eff}(dc)$ versus frequency, microstrip line circuit model.

nately, k must be found either from a rigorous analysis of the particular frequency-dependent line, or by an experimental measurement, for a given line.

Rewriting (5.28) in terms of the effective dielectric constant,

$$\frac{\epsilon_{eff}(\omega)}{\epsilon_{eff}(\omega = 0)} = \frac{\epsilon_{eff}(\omega)}{v_0^2 LC} = 1 + \frac{\sqrt{k^4 + (2\omega^2 LC_m)^2} - k^2}{2\omega^2 LC} \tag{5.33}$$

Figure 29 gives a set of plots of (5.33), for several different values of the parameters as shown versus frequency. Clearly there is a range of frequencies that can be called the quasi-static region. Note also that ϵ_r is not a rapidly changing function of frequency for most practical cases in the frequency-dependent range, the microstrip line can be treated as an ordinary TEM system. The line parameters of course must be properly calculated at the operating frequency.

5.3 THE MICROSTRIP SLOW WAVE MODE

When semiconductor devices are designed to be used as amplifiers at UHF and at higher frequencies, it is convenient to include the input and output connecting lines as microstrip lines fabricated on the semiconductor itself. Figure 30 illustrates a typical system. A silicon wafer with a metallized bottom is subjected to an oxidation process that converts a thin layer on the top surface of the wafer to the insulator silicon dioxide. The microstrip upper conductor is then fabricated (usually thin film) on the silicon dioxide.

Although pure or nearly pure silicon has a high enough resistivity to form a fairly low loss dielectric for a microstrip line, the resistivity of the silicon wafer used in semiconductor device manufacture is usually dictated by the needs of the semiconductor device designer. In this case it is important that the path through the silicon to the lower metallization have a reasonably low resistance; consequently the silicon itself cannot be considered to be a good insulator. The silicon dioxide layer is grown on the surface of the silicon to increase the resistance to ground seen by the upper conductor of the microstrip line.

At low frequencies this microstrip system propagates signals in the usual quasi-TEM mode. Because of the silicon–silicon dioxide double layer dielectric, however, there may exist a slow wave mode over a certain frequency range. Although the compound dielectric is made up of materials with an $\epsilon_r \simeq 12$ (silicon) and $\epsilon_r \approx 4$ (silicon dioxide), the slow wave mode may propagate as slowly as if the dielectric had an ϵ_r of 1600.

The quasi-TEM (quasi-static) mode in this system is characterized by the silicon behaving as a dielectric, and ϵ_r for the entire dielectric is essentially 12. At high enough frequencies the silicon acts as a lossy metal (see the skin effect discussions in Chapter 7) and the effective dielectric is the thin silicon dioxide layer. The wave propagation in this latter case is essentially a surface wave in the silicon dioxide layer.

Between the two above-mentioned regions (in frequency) a wave may be launched in the silicon layer that essentially "ricochets" between the upper and lower boundaries of the silicon layer. Since the resulting propagation speed along the surface is very slow, an extremely high effective dielectric constant may be defined. The wave is propagating through a lossy material; therefore the slow wave mode is characterized by a high attenuation per unit length.

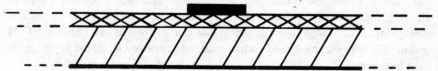

Figure 30 Cross section of a line that may exhibit the microstrip slow wave mode.

The surface wave mode dielectric thickness is usually in the order of 5000 Å. Because of this, the frequencies at which the frequency-dependent behavior as described in Section 5.2 become noticeable are so high that they are not of practical interest. The high frequency mode for a double layer dielectric system that can exhibit a slow wave mode is therefore a second quasi-static mode.

5.4 THE PARALLEL STRIP LINE ABOUT A SHEET OF DIELECTRIC

The transmission line formed by a pair of parallel conducting strips in a uniform dielectric space is a conventional TEM line. If the region between the planes of the conductors is a dielectric that is different from the region outside these planes, a balanced line similar in properties to microstrip is formed (Figure 31).

Because of the symmetry about the y axis of the line in Figure 31, if the line is driven in a balanced mode, the y axis is at ground potential. The electric field lines above the y axis are therefore identical to those of the microstrip line formed by the system above the y axis, and the electric field lines below the y axis are the mirror image (about the y axis) of these above. By symmetry, therefore, $C_{bal} = \frac{1}{2}C_{\mu strip}$, and $L_{bal} = 2L_{\mu strip}$. The formulas of Section 5.1 apply to finding these values, and $Z_{0(bal)} = Z_{0(\mu strip)}$ for the case when the balanced line can be thought of as a microstrip line plus its mirror image about the ground plane (see Figure 3).

For dielectric materials with $\epsilon_r \gtrsim 3$, the electric field lines will not extend far in the x direction beyond the metal strips. The formula above therefore will be accurate even if the dielectric extends only several conductor widths in the x direction. This type of structure is very easy to manufacture accurately, and it is a good means of realizing a low-Z_0 balanced line that is flexible and easy to handle.

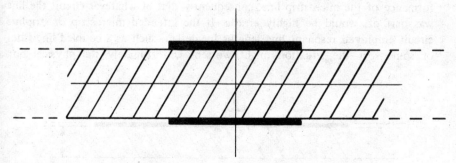

Figure 31 Two-sided parallel strip line about a dielectric.

5.5 SLOTLINE

When two thin conducting half-planes are placed edge to edge in the same plane, the resulting line is TEM if the dielectric cross section is uniform. When the lines are placed on one side of a dielectric slab, as in Figure 32 the result is a non-TEM line called slotline. Slotline is an unusual transmission line in that a principal dimension—the gap between the planes—refers to the only place where no conductor is present, rather than to an actual conductor dimension. Thus it is not possible to design a "connector" of any sort to ideally interface slotline with other types of transmission line. Typically, slotline signals are "launched" by a coaxial cable interconnection (Figure 33a) or by building a hybrid microstrip-slotline system with a transition (Figure 33b).

Interestingly enough, most transmission line devices such as directional couplers and coupled line filters can be built using slotline by simply constructing the photographic "negative" of the microstrip version of the network. The actual device dimensions may have to be adjusted for the very different mode of operation.

Figure 34 shows Z_0 versus the ratio of the conductor gap to the dielectric thickness for slotline, based on a dc calculation, for $\epsilon_r = 9.6$. The usefullness of this curve for actual rf frequencies is demonstrated in Figure 35, where the characteristic impedance of lines of several different widths is plotted versus frequency.

The importance of slotline is perhaps not as a practical transmission line system, but as a transmission line that is often inadvertently built into microstrip or stripline networks and acts as a very erratic flaw condition. For example, consider the microstrip line shown in Figure 36. Poor construction techniques have caused a small gap in the ground plane where several metal sections are joined. Comparing Figure 36 to Figure 33b, we see that this microstrip line has built into it a slotline resonant section and a coupling mechanism. By design, this circuit would be called a bandstop filter. By accident, the response would have an annoying "suckout." In a production environment where the gap in the microstrip ground plane would be unpredictable in width or length, the performance of the microstrip line, consequently that of whatever circuit the line was part of, would be highly erratic. If the intended microstrip or stripline circuit employed resonant line lengths by design—such as a coupled line filter of some sort—the situation might worsen. Unfortunately the physical box

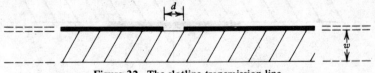

Figure 32 The slotline transmission line.

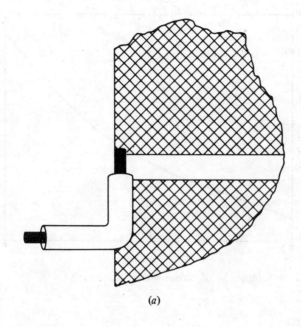

(a)

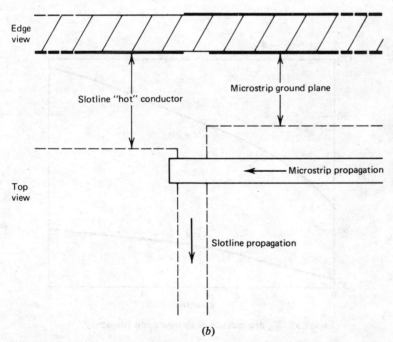

Edge view

Slotline "hot" conductor

Microstrip ground plane

Top view

Microstrip propagation

Slotline propagation

(b)

Figure 33 (a) Slotline–coaxial cable connection. (b) Slotline-microstrip interface.

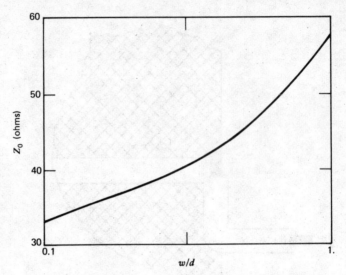

Figure 34 Z_0 versus w/d for slotline at dc.

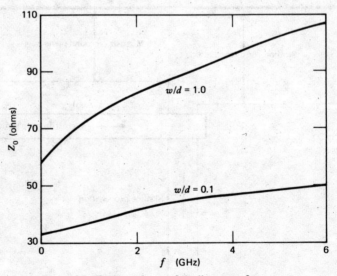

Figure 35 Z_0 dependence of slotline upon frequency.

78

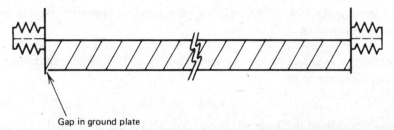

Gap in ground plate

Figure 36 Microstrip circuit with a parasitic slotline coupling.

dimensions and ground plane lengths in microstrip and stripline circuits are often close to integral multiples of $\frac{1}{4}$ wavelengths at the operating frequency of the device. The unwanted slotline resonances caused by poor design and/or construction then appear (with statistical regularity) at the frequencies where they can do the most damage to the intended circuit performance.

5.6 HELICAL LINE

A transmission line can be formed by coiling a wire in the form of a helix, as in Figure 37. This type of structure has two principal applications:

1. A transmission line resonator may be shortened without serious degradation of Q, provided the shortening is not drastic. This is often necessary to meet practical packaging and size requirements for electronic instruments of various kinds.

2. Assuming that the traveling wave follows the wire at the speed of light, the actual wave velocity in the z direction must be much less than the speed of light. This type of structure is therefore sometimes called a slow wave structure. If an electron beam were to be passed along the axis of the helix, the velocity of the beam could be matched to the wave velocity of the transmission line (in the z direction) and energy could be transferred between them. This energy transfer

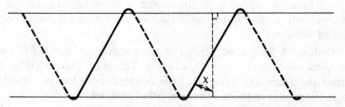

Figure 37 The helical transmission line.

mechanism is the basis of the traveling wave tube, an important type of microwave power amplifier.

Intuitively, one would expect the wave velocity of the helical transmission line in the z direction to be

$$v \simeq C_0 \sin \chi \qquad (5.34)$$

where χ is defined in Figure 37. For χ small, (5.34) is quite accurate. In general, (5.34) must be modified to take into account the complex interactions of the fields due to the spiral helix boundary conditions. The full analysis is approximated by replacing the helical coil with a hypothetical current carrying anisotropic sheet that allows current flow only along a spiral path. This analysis is beyond the scope of this treatment, and the reader is referred to the materials listed below for further information.

5.7 SUGGESTED FURTHER READING

1. H. A. Wheeler, "Transmission-Line Properties of Parallel Strips Separated by a Dielectric Sheet," *IEEE Transactions on Microwave Theory and Techniques*, Vol. MTT-12, May 1964. This derivation shows an elegant use of conformal transformations and mathematical approximations to yield the quasi-static formulas for unshielded microstrip line.

2. H. A. Wheeler, "Transmission-Line Properties of a Strip on a Dielectric Sheet on a Plane," *IEEE Transactions on Microwave Theory and Techniques*, Vol. MTT-25, August 1977. Nearly 15 years after writing the paper cited in Reference 1, Dr. Wheeler reviewed his earlier work and derived several more exact and easier-to-use explicit formulas.

3. R. Mittra and T. Itoh, "Analysis of Microstrip Transmission Lines," in *Advances in Microwaves*, Vol. 8, L. Young and H. Sobol, Editors, Academic Press, New York, 1974. A very thorough treatment of several techniques of analyzing microstrip for both the quasi-static and frequency-dispersive mode.

4. W. Gettsinger, "Microstrip Dispersion Model," *IEEE Transactions on Microwave Theory and Techniques*, Vol. MTT-21, January 1973. Describes a very straightforward model that allows for a determination of the necessary parameter to complete the model shown in Section 5.4, based on the geometry of the line.

5. H. Carlin, "A Simplified Circuit Model for Microstrip," *IEEE Transactions on Microwave Theory and Techniques*, Vol. MTT-21, September 1973. Discusses the origin and validity of the coupled mode circuit model and compares the results to those of several other analyses.

6. M. Schneider, "Microstrip Dispersion," *Proceedings of the IEEE*, Vol. 60, No. 1, January 1972. Presents an empirical formula for the frequency-

dependent characteristic of microstrip that is quite accurate and simple. Of particular interest is the approximation that $\partial^2 v/\partial\omega^2 = 0$ [see (5.28)] near the cutoff frequency of the lowest order transverse electric surface wave mode. This frequency is given in terms of the dielectric thickness h as follows:

$$\omega_C = \frac{\pi v_0}{2h\sqrt{\epsilon_r - 1}}$$

Although a justification of these statements is well outside the scope of this treatment, the use of them provides the missing parameter for a complete formula for the frequency-dependent properties of the microstrip line in Section 5.2.

7. H. Hasegawa et. al., "Properties of Microstrip Line on Si–SiO$_2$ System," *IEEE Transactions on Microwave Theory and Techniques*, Vol. MTT-19, No. 11, November 1971. Derives the basic relations for the slow wave microstrip mode.

8. E. Mariani et al., "Slot Line Characteristics," *IEEE Transactions on Microwave Theory and Techniques*, Vol. MTT-17, No. 12, December 1969. Derives the basic properties of slotline and gives several curves of effective dielectric constant and characteristic impedance versus frequency. Has sketches of basic launching techniques and some simple slotline networks.

9. S. Cohn, "Sandwich Slot Line," *IEEE Transactions on Microwave Theory and Techniques*, Vol. MTT-19, No. 9, September 1971. If a slotline is formed with dielectric sheets placed symmetrically top and bottom, the resulting system is referred to as a sandwich slot line. Cohn gives the basic relations for this type of line.

10. J. R. Pierce, *Travelling-Wave Tubes*, Van Nostrand, New York, 1950. Treats the current-sheet formulation of the helical transmission line problem.

6

An Introduction to Conformal Transformations

Conformal transformation is a mathematical technique that allows a particular transmission line geometry to be transformed into a new geometry in a second coordinate system, with certain rules governing the relationship between the electrical properties of the lines in the two systems. If the second system is judiciously chosen, the new geometry is more amenable to being solved by Laplace's equation (in the plane) than was the original geometry.

The theory of conformal transformations is a topic in the more general subject of the theory of complex variables. Certain proofs that would require lengthy discussions of complex variable theory are postulated. The treatments in the references at the end of the chapter contain all the proofs required to make the derivation rigorous.

Following the derivation, several examples are worked out. These are practical transmission line cases that may or may not be treatable by other means. Two important problems have been solved analytically by conformal transformations only—stripline and microstrip. Both the latter examples are quite lengthy and require a concerted study of the topic for full appreciation. The treatment of this chapter should be adequate to allow an appreciation of the work done, and to provide a reasonable basis for a more detailed study if so desired.

6.1 CONFORMAL TRANSFORMATIONS

Let us consider a transmission line cross section that is defined by two equipotential surfaces S_1 and S_2 in the x-y plane. To find the line parameters, we shall solve Laplace's equation in the plane,

$$\nabla^2 \varphi = \frac{\partial^2 \varphi}{\partial x^2} + \frac{\partial^2 \varphi}{\partial y^2} = 0 \tag{6.1}$$

82

subject to the boundary conditions

$$\varphi = \varphi_1 \quad \text{on } S_1 \tag{6.2a}$$

$$\varphi = \varphi_2 \quad \text{on } S_2 \tag{6.2b}$$

It is desirable to find a new coordinate system, the (u, v) system, in which the problem is easily solvable. In general,

$$u = u(x, y) \tag{6.3}$$

$$v = v(x, y) \tag{6.4}$$

Consider for the moment the complex transformation

$$W(u, v) = u + jv = F(Z) \tag{6.5}$$

where

$$Z = x + jy = Re^{j\theta} \tag{6.6}$$

$$x = R \cos(\theta)$$

$$y = R \sin(\theta)$$

As an example of such a function, let

$$W = \text{Ln}(Z) = \text{Ln}(Re^{j\theta}) = \text{Ln}(R) + j\theta \tag{6.7a}$$

from which

$$u = \text{Ln}(R) \tag{6.7b}$$

and

$$v = \theta \tag{6.7c}$$

Assuming continuous and "well-behaved" functions, every point in the Z plane will correspond to some point in the W plane. If the curve in the Z plane is closed, its mapping the W plane is closed.* In the example above, a circle of constant radius R_0 in the xy (Z) plane maps into a vertical line at $u = \text{Ln}(R_0)$ in the $uv(W)$ plane as in Figure 38. Also, a radial line at angle θ_0 in the xy plane maps into a horizontal line at $v = \theta_0$ in the uv plane, as in Figure 39.

A function $F(Z)$ is defined to be analytic at some point Z_0 if the derivative

$$\frac{dw}{dZ} = \frac{dF(Z)}{dZ} = \lim_{\Delta z \to 0} \frac{F(Z + \Delta Z) - F(Z)}{\Delta Z} \tag{6.8}$$

exists and is unique at Z_0 regardless of the path ΔZ. If $F(Z)$ is analytic over some region of interest, the transformation, or mapping, is called conformal.

---

*A closed curve in the complex plane sometimes goes to infinity, then returns "from" infinity at a point 180 degrees away from the departure point.

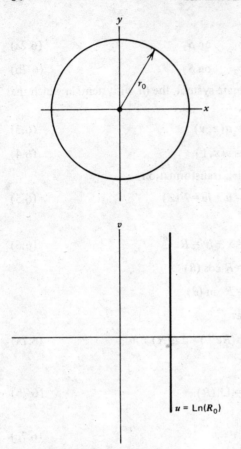

$u = \mathrm{Ln}(R_0)$

Figure 38 The function $W = \mathrm{Ln}(z)$ mapping of the circle of radius R_0.

Now, consider two different paths for ΔZ:

(a) $\Delta Z = \Delta x$ from which

$$\frac{dW}{dZ} = \frac{dW}{dx} = \frac{\partial u}{\partial x} + j\frac{\partial v}{\partial x} \tag{6.9}$$

(b) $\Delta Z = j\Delta y$ from which

$$\frac{dW}{dZ} = \frac{dW}{jdy} = -j\frac{dW}{dy} = -j\left(\frac{\partial u}{\partial y} + j\frac{\partial v}{\partial y}\right) = \frac{\partial v}{\partial y} - j\frac{\partial u}{\partial y} \tag{6.10}$$

For an analytic function, the two cases must be equal,

$$\frac{\partial u}{\partial x} = \frac{\partial v}{\partial y} \tag{6.11a}$$

$$\frac{\partial v}{\partial x} = \frac{-\partial u}{\partial y} \tag{6.11b}$$

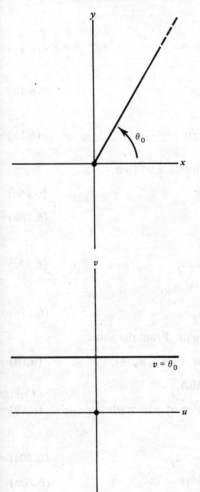

Figure 39 The function $W = \text{Ln}(z)$ mapping of a vector at an angle θ_0.

Equations 6.11a and 6.11b are known as the Cauchy-Riemann conditions. Returning to the example, (6.7), we can write

$$\frac{\partial u}{\partial x} = \frac{x}{x^2 + y^2} = \frac{\partial v}{\partial y} \tag{6.12a}$$

$$\frac{\partial u}{\partial y} = \frac{y}{x^2 + y^2} = \frac{-\partial v}{\partial x} \tag{6.12b}$$

For this example, the Cauchy-Riemann conditions are satisfied. Note, however, that $F = \text{Ln}(Z)$ is not analytic at $R = 0$.

There are several easily derived properties of conformal transformations that will be necessary in the derivations that follow. For convenience, we use the shorthand notation for partial derivatives,

$$u_x \equiv \frac{\partial u}{\partial x} \tag{6.13}$$

$$u_{xx} = \frac{\partial^2 u}{\partial x^2} \tag{6.14}$$

$$u_{xy} = \frac{\partial^2 u}{\partial x \partial y}, \text{ etc.} \tag{6.15}$$

In this notation, the Cauchy-Riemann conditions are written

$$u_x = v_y \tag{6.16a}$$

$$v_x = -u_y \tag{6.16b}$$

Now,

$$\nabla u = \hat{a}_x u_x + \hat{a}_y u_y \tag{6.17a}$$

and

$$\nabla v = \hat{a}_x v_x + \hat{a}_y v_y \tag{6.17b}$$

where $\hat{a}_x, \hat{a}_y$ are unit vectors in the (x, y) system. From the above,

$$\nabla u \cdot \nabla v = u_x v_x + u_y v_y = u_x v_x - v_x u_x = 0 \tag{6.18}$$

The (u, v) system is therefore orthogonal. Also,

$$|\nabla u|^2 \equiv u_x^2 + u_y^2 = v_y^2 + (-v_x)^2 = v_y^2 + v_x^2 = |\nabla v|^2 \tag{6.19}$$

and

$$u_{xx} = v_{yx} = v_{xy} = -u_{yy} \tag{6.20a}$$

$$\therefore u_{xx} + u_{yy} = 0 \tag{6.20b}$$

and similarly,

$$\nabla^2 v = v_{xx} + v_{yy} = 0 \tag{6.20c}$$

The usefulness of conformal transformations to the (transmission line) problem becomes apparent when we calculate $\nabla^2 \varphi$ in the (u, v) system.* By the usual chain rules of differentiation,

$$\varphi_x = \varphi_u u_x + \varphi_v v_x \tag{6.21a}$$

$$\varphi_y = \varphi_u u_y + \varphi_v v_y \tag{6.21b}$$

---

*This derivation is handled very elegantly through the use of general curvilinear coordinate notation. See, for example, Reference 1 at the end of this chapter.

and

$$\varphi_{xx} = (\varphi_{ux}u_x + \varphi_u u_{xx}) + (\varphi_{vx}v_x + \varphi_v v_{xx}) \qquad (6.22a)$$

$$\varphi_{yy} = (\varphi_{uy}u_y + \varphi_u u_{yy}) + (\varphi_{vy}v_y + \varphi_v v_{yy}) \qquad (6.22b)$$

Using the above and (6.20) yields

$$\nabla^2_{xy}\varphi \equiv \varphi_{xx} + \varphi_{yy} = \varphi_{ux}u_x + \varphi_{uy}u_y + \varphi_{vx}v_x + \varphi_{vy}v_y \qquad (6.23)$$

Furthermore, using (6.16) we have

$$\varphi_{xx} + \varphi_{yy} = u_x(\varphi_{ux} + \varphi_{vy}) + u_y(\varphi_{uy} - \varphi_{vx}) \qquad (6.24)$$

Again, using the chain rule gives

$$\varphi_{xx} + \varphi_{yy} = u_x[\varphi_{uu}u_x + \varphi_{uv}v_x + \varphi_{vu}u_y + \varphi_{vv}v_y]$$
$$+ u_y[\varphi_{uu}u_y + \varphi_{uv}v_y - \varphi_{vu}u_x - \varphi_{vv}v_x] \qquad (6.25)$$

and after again using (6.16),

$$\varphi_{xx} + \varphi_{yy} = \varphi_{uu}u_x^2 + \varphi_{vv}u_x^2 + \varphi_{uu}u_y^2 + \varphi_{vv}u_y^2 = (\varphi_{uu} + \varphi_{vv})(u_x^2 + u_y^2)$$
$$(6.26)$$

Since φ satisfies Laplace's equation in the (x, y) system and, for a nontrivial transformation, $u_x^2 + u_y^2 \neq 0$,

$$\varphi_{xx} + \varphi_{yy} = \varphi_{uu} + \varphi_{vv} = 0 \qquad (6.27)$$

In other words, we are guaranteed a priori by the properties of conformal transformations that φ will satisfy Laplace's equation in the (u, v) system. Note that this guarantee holds for any choice of transformation, provided it is well behaved enough to allow the operations of the derivation above.

Let us calculate the energy stored in the electric field in both coordinate systems. In the (x, y) system this is simply

$$\tfrac{1}{2}CV^2 = \tfrac{1}{2}C(\varphi_2 - \varphi_1)^2 = \frac{\epsilon}{2}\iint\limits_{\substack{x-y \\ \text{plane}}} (\varphi_x^2 + \varphi_y^2)\,dx\,dy \qquad (6.28)$$

Consider first the integrand in (6.28):

$$\varphi_x^2 + \varphi_y^2 = [\varphi_u u_x + \varphi_v v_x]^2 + [\varphi_u u_y + \varphi_v v_y]^2 \qquad (6.29a)$$

where (6.21) was used above. Again, using (6.16), we write

$$\varphi_x^2 + \varphi_y^2 = \varphi_u^2 u_x^2 + \varphi_v^2 v_x^2 + \varphi_v^2 v_y^2 + \varphi_u^2 u_y^2 \qquad (6.29b)$$

$$= \varphi_u^2(u_x^2 + u_y^2) + \varphi_v^2(v_x^2 + v_y^2) \qquad (6.29c)$$

and using (6.19),

$$\varphi_x^2 + \varphi_y^2 = (\varphi_u^2 + \varphi_v^2)(u_x^2 + u_y^2) \tag{6.29d}$$

The differential area $dx\, dy$ is transformed according to

$$dx\, dy = J \left| \frac{x, y}{u, v} \right| du\, dv = \frac{1}{u_x v_y - v_x u_y} du\, dv = \frac{du\, dv}{u_x^2 + u_y^2} \tag{6.30}$$

where J is the Jacobian of the transformation.

Combining the expression for the integrand (6.29) and the expression for the differential area (6.30), we get

$$\tfrac{1}{2}\, \epsilon \iint_{\substack{x-y \\ \text{plane}}} (\varphi_x^2 + \varphi_y^2)\, dx\, dy = \tfrac{1}{2}\, \epsilon \iint_{\substack{u-v \\ \text{plane}}} (\varphi_u^2 + \varphi_v^2)\, du\, dv \tag{6.31}$$

Equations 6.28 and 6.31 show that the capacitance (per unit length) of the structure in the (u, v) system is identical to the original capacitance, calculated in the (x, y) system. Since the transformation concerned itself only with the transmission line cross section, the length dimension is unaffected, and "per unit length" considerations are unaltered.

The net result of these derivations is this: for a given transmission line geometry, if a transformation can be found to a (u, v) system in which the capacitance of the resulting structure can be calculated, the problem is solved. Note that this assurance lends no help whatsoever in finding a reasonable transformation. There also exists the possibility of multiple transformations—a transformation of a transformation (etc.) will still yield the same capacitance. In practice, the procedure is similar in intent to the transformation of variables techniques of elementary integral calculus. That is, there is no guarantee that a given transformation will be of any use, but it might lead to another transformation that will render the problem solvable.

6.2 EXAMPLE: THE COAXIAL CABLE

Consider a coaxial cable, as was shown in cross section in Figure 2, Chapter 1. The inner conductor radius is a, the outer conductor radius is b. The inner conductor is at potential φ_1 and the outer conductor is at potential φ_2. Repeating the transformation example of the previous section, let

$$W = \text{Ln}(Z) = \text{Ln}(r) + j\theta \tag{6.32a}$$

from which

$$u = \text{Ln}(r) \tag{6.32b}$$

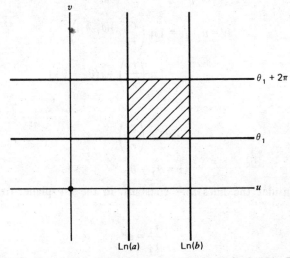

Figure 40 The coaxial cable cross section transformed using $W = \text{Ln}(z)$.

and

$$v = \theta \qquad (6.32c)$$

The equipotential surfaces $r = a$ and $r = b$ transform to lines of constant u, $u = \text{Ln}(a)$ and $u = \text{Ln}(b)$, for all θ, as shown in Figure 40.

The capacitance of the coaxial cable is defined by any region of θ, $\theta_1 \leqslant \theta \leqslant \theta_1 + 2\pi$. Since the equipotential lines extend from $-\infty$ to $+\infty$, the field lines are uniform and parallel, and the capacitance is that of a section of an ideal parallel plate capacitor,

$$C = \frac{\epsilon(2\pi)}{\text{Ln}(b) - \text{Ln}(a)} = \frac{2\pi\epsilon}{\text{Ln}(b/a)} \qquad (6.33)$$

6.3 EXAMPLE: THE PARALLEL WIRE LINE

Consider the transformation

$$W = \text{Ln}\left(\frac{Z - a}{Z + a}\right) \qquad (6.34)$$

where a is real and positive.
Let

$$Z - a = re^{j\theta_1} \qquad (6.35a)$$

$$Z + a = re^{j\theta_2} \qquad (6.35b)$$

Then

$$W = u + jv = \text{Ln}\left(\frac{r_1}{r_2} e^{j(\theta_1 - \theta_2)}\right) \tag{6.36}$$

$$= \text{Ln}\left(\frac{r_1}{r_2}\right) + j(\theta_1 - \theta_2)$$

and

$$u = \text{Ln}\left(\frac{r_1}{r_2}\right) \tag{6.37}$$

$$v = \theta_1 - \theta_2 \tag{6.38}$$

Let us consider the lines $u =$ constant. In the x-y plane, from (6.37), we know that

$$\frac{r_1}{r_2} = e^u \tag{6.39}$$

or

$$\frac{r_1^2}{r_2^2} = e^{2u} = \frac{(x-a)^2 + y^2}{(x+a)^2 + y^2} \tag{6.40}$$

Rearranging this equation, equivalently,

$$(x - a\coth(u))^2 + y^2 = (a\,\text{csch}(u))^2 \tag{6.41}$$

Consider now a circle of radius b, located with a center at $(c, 0)$ (Figure 41). By direct comparison,

$$a\coth(u) = c \tag{6.42a}$$

$$a\,\text{csch}(u) = b \tag{6.42b}$$

Dividing (6.42b) by (6.42a) yields

$$\frac{\coth(u)}{\text{csch}(u)} = \frac{c}{b} \tag{6.43}$$

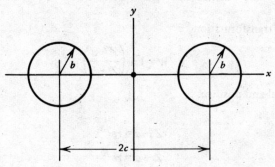

Figure 41   Cross section of a parallel round wire transmission line.

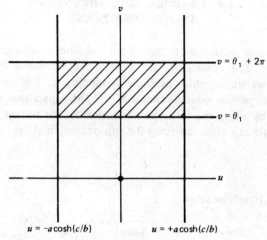

Figure 42 The transformed parallel round wire transmission line.

or

$$\frac{c}{b} = \cosh(u) \tag{6.44}$$

and finally

$$u = \cosh^{-1}\left(\frac{c}{b}\right) \tag{6.45}$$

Similarly, a circle of radius b with a center at $(-c, 0)$ will yield

$$u = -\cosh^{-1}\left(\frac{c}{b}\right) \tag{6.46}$$

If the two circles in the x-y plane represent the cross section of a parallel wire transmission line, the transformation (6.34) has transformed this line geometry into that of two equipotential lines of constant u (Figure 42). As in the previous example, the capacitance is defined by a 2π region in $\theta_2 - \theta_1$, which by (6.38) becomes $0 \leqslant v \leqslant 2\pi$. The capacitance is then simply

$$C = \frac{\epsilon(2\pi)}{\cosh^{-1}(c/b) - (-)\cosh^{-1}(c/b)} = \frac{\epsilon\pi}{\cosh^{-1}(c/b)} \tag{6.47}$$

Section 5.1 used the method of images to extend the solution for a narrow strip conductor above a ground plane to hold for two narrow parallel strip conductors. By the same reasoning, the solution above for two parallel circular conductors can be extended to hold for one circular conductor above a ground plane. This is done by merely noting that the plane of symmetry between the two circular conductors must be an equipotential surface.

6.4 EXAMPLE: TWO THIN STRIPS IN THE SAME PLANE

Consider the cross section of a pair of parallel thin conductors, in a uniform dielectric space. The conductors are a wide and separated by a distance d. Figure 43 shows the location of these conductors in the x-y plane. Assume for the purposes of this example that $a \gg d$. The capacitance of this structure can be found by using the formulas of Section 5.1, or noting that in the case of a parallel plate capacitor satisfying the approximation above,

$$C \simeq \frac{\epsilon a}{d} \tag{6.48}$$

Consider the transformation

$$W = e^{\pi Z/d} = e^{\pi x/d} \left[\cos\left(\frac{\pi y}{d}\right) + j \sin\left(\frac{\pi y}{d}\right) \right] \tag{6.49}$$

The lower conductor of the structure in Figure 43, at $y = 0$, transforms according to

$$v = 0 \tag{6.50a}$$

$$u = e^{\pi x/d} \tag{6.50b}$$

For $-a/2 \leqslant x \leqslant a/2$, $e^{-\pi a/2d} \leqslant u \leqslant e^{\pi a/2d}$. Figure 44 gives this section of line. Similarly, the upper conductor, at $y = d$ in Figure 43, transforms to the line $v = 0$, $-e^{\pi a/2d} \leqslant u \leqslant -e^{-\pi a/2d}$, also shown in Figure 44.

It can be seen by examining Figure 44 that if a potential difference exists between the two lines, most of the charge will reside on the lines close to the gap between the two lines. From the figure, this gap is of length

$$g = 2e^{-\pi a/2d} \tag{6.51}$$

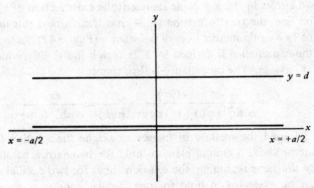

Figure 43 The parallel plate capacitor (cross section).

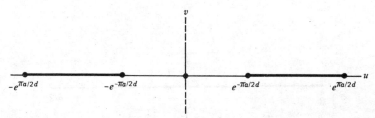

Figure 44 The edge-to-edge plate capacitor obtained by transformation from the parallel plate capacitor.

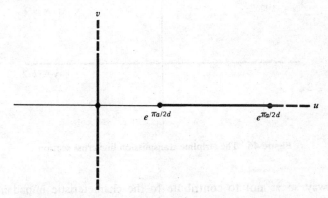

Figure 45 The plate—normal to a metal ground plane capacitor.

Since $a \gg d$, $e^{\pi a/2d}$ is so much greater than $e^{-\pi a/2d}$ that the edges of the lines far away from the gap play little part in determining the capacitance between the lines. Their exact location is therefore of little importance, and the lines can be considered to be infinitely wide. Inverting (6.51) and substituting into (6.48), then, we have

$$C \cong \epsilon \left[\frac{2}{\pi} \mathrm{Ln} \left(\frac{2}{g} \right) \right] \qquad (6.52)$$

The method of images again predicts a second use for a capacitance formula, as shown in Figure 45. Equation 6.52 will give the capacitance of a wide, thin plate held normal to an infinite plane, with a gap of $g/2$.

6.6 STRIPLINE

Figure 46 shows what is usually called a stripline or simply a strip transmission line. A thin, narrow strip of conductor is embedded between two grounded conductors. The most common case is that of a centered strip, with side walls far

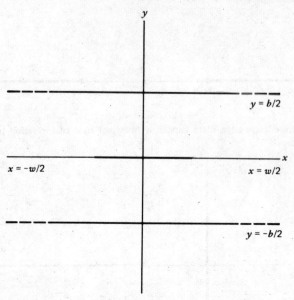

Figure 46   The stripline transmission line cross section.

enough away so as not to contribute to the characteristic impedance of the system. The more general cases have been treated and are referenced at the end of this chapter; they are not considered here. Since all electric and magnetic fields are contained between the outer planes, the stripline is a TEM system regardless of what dielectric the line is filled with, as long as the space between the outer planes is uniformly filled with dielectric.

If the center conductor of a centered stripline, (referred to simply as a stripline here after) were to be replaced with a round wire of diameter d, the capacitance, hence the characteristic impedance of the line created, could be solved analytically by a conformal transformation. Without working through the details, the desired result is

$$Z_0 = \frac{60}{\sqrt{\epsilon_r}} \, \text{Ln} \left(\frac{4b}{\pi d} \right) \tag{6.53}$$

Again, by the technique of conformal transformations, an equivalence can be established between the round center conductor and a thin, narrow strip center conductor:

$$d = \frac{w}{2} \left[1 + \frac{t}{w} \left(1 + \text{Ln} \, \frac{4\pi w}{t} + 0.51\pi \left(\frac{t}{w} \right)^2 \right] \tag{6.54}$$

The equations above are approximate and are only reasonable for lines satisfying $w < 0.3b$. Very accurate approximations for all possible ranges of the parameters have been derived and are in the literature. The references at the end of this chapter cover several of these.

Stripline circuits are important practically because of the ease of circuit fabrication that is inherent in the nature of stripline. Typically, a convenient dielectric sheet is copper clad on both sides, and the desired stripline circuit is photoetched using conventional printed circuit techniques on one side. A second dielectric sheet (identical to the first) is copper clad on only one side and is then bolted or somehow clamped to the first sheet, with the unclad section of the second sheet in intimate contact with the circuit side of the first sheet. As a rule the edges are "sealed," since the structure, unsealed, would act as a cutoff wave guide and radiate power at the edges.

The stripline circuit has several advantages over the microstrip circuit. First, since the line is a true TEM line and there is no effective dielectric correction factor, stripline circuits tend to be smaller than their microstrip counterparts. This size advantage also pertains to the height, or thickness, direction since in stripline there is never a need to keep a shielding cover "far away" from the line. Broadband circuits are easy to fabricate, easy to duplicate, and inexpensive.

On the other hand, stripline circuits have found relatively little use as narrow band circuits. This is because the process of joining the two sections together is hard to reproduce exactly, and, once joined, the stripline circuit is very hard to adjust or "trim."

6.7 SUGGESTED FURTHER READING

1. R. E. Collin, *Field Theory of Guided Waves*, McGraw-Hill, New York, 1960. A very thorough and rigorous treatment of conformal mappings, including several topics not mentioned in the introductory treatment in this chapter. Also gives a rigorous derivation of the stripline characteristic impedance.

2. M. V. Schneider, "Microstrip Lines for Microwave Integrated Circuits," *Bell System Technical Journal*, May–June, 1969. Analytically solves the thin conductor microstrip (quasi-static) problem exactly. The resulting formulas are not convenient for day-to-day use, so approximations are given that agree well with those presented here. The exercise in conformal transformations, however, is excellent.

3. W. R. Smythe, *Static and Dynamic Electricity*, McGraw-Hill, New York, 1950. Contains a plethora of examples of transmission line parameters found by conformal mapping techniques.

4. H. Howe, Jr., *Stripline Circuit Design*, Artech House, Dedham, Mass., 1974. Gives formulas and tables for off-centered as well as centered striplines in various dielectrics. As the title suggests, the entire topic of stripline circuits is covered in detail.

5. J. C. Tippet and D. C. Chang, "Radiation Characteristics of Dipole Sources
   Located Inside a Rectangular, Coaxial Transmission Line," National Bureau
   of Standards Industrial Report 75-829, available from National Technical
   Information Service, Springfield, Va. Makes use of the characteristic impedance formula for a totally enclosed stripline, taking the location of the
   sidewalls into account. The derivation of this formula, by conformal mapping
   techniques, is contained in an appendix.

7

The Skin Effect and Losses
in Transmission Lines

Losses in transmission lines can be grouped into three broad categories. These categories are metallic (ohmic) losses, dielectric losses, and spurious radiation. Of these three mechanisms, ohmic loss is the only one that lends itself to detailed analysis. Radiation losses occur at imperfections and discontinuities in transmission lines. In many cases a line that is not balanced (or unbalanced) properly has return currents flowing through some unspecified path that is not enclosed or enclosing and is not unidimensional. The line then acts partially as an antenna, and energy is lost from the line. Dielectric losses occur because of energy dissipation in the dielectric material itself; these are summarized in the loss parameter ϵ', as shown in Chapter 1. Although dielectric losses are sometimes predominant in a transmission line system, the only answer to the problem is to find a less lossy dielectric material. The physical mechanisms of dielectric loss require detailed study of the dielectric material itself—the dielectric often being some organic plastic or ceramic material. This type of material study is difficult and material specific, and is beyond the scope of this treatment.

Ohmic losses in transmission lines occur in the metallic conductors that make up the transmission lines. Since these metals have nonzero resistivities, the electric and magnetic fields that carry the propagating energy penetrate the metals somewhat, and there is heating of the metals. The penetration of rf fields into the surface of metals is known as the *skin effect*. The amount of loss per unit length in a line is a function of the metals used and the geometry of the line itself. The line geometry contributes to loss calculations because of a second effect, sometimes called the proximity effect. The proximity effect is a tendency of the currents in a transmission line to concentrate in the regions of the conductors where the magnetic fields that produced the currents are strongest. If the current is bunched into a relatively small fraction of the available metal cross section, the line will be lossy, whatever the composition of the conductor itself.

7.1 THE SKIN EFFECT EQUATION

The basic phenomenon of fields penetrating into real metals can be examined by deriving a differential equation for the electric and magnetic fields in the metals, called the skin effect equation. From Maxwell's equations,

$$\nabla \times \mathbf{E} = -j\omega\mu\mathbf{H} \tag{7.1}$$

and by vector identity,

$$\nabla \times \nabla \times \mathbf{E} = \nabla (\nabla \cdot \mathbf{E}) - j\omega\mu \nabla \times \mathbf{H} \tag{7.2}$$

In a good conductor it is reasonable to assume that there is no space charge. That is,

$$\nabla \cdot \mathbf{D} = \epsilon\nabla \cdot \mathbf{E} = 0 \tag{7.3}$$

and that electron current will dominate. In other words, displacement current may be ignored. Therefore,

$$\nabla \times \mathbf{H} = \mathbf{J} + j\omega\epsilon\mathbf{E} \simeq \mathbf{J} \tag{7.4}$$

Substituting (7.3) and (7.4) into (7.2),

$$\nabla^2 \mathbf{E} = j\omega\mu\mathbf{J} \tag{7.5}$$

By Ohm's law, the foregoing also can be written

$$\nabla^2 \mathbf{J} = j\omega\mu\sigma\mathbf{J} \tag{7.6}$$

or

$$\nabla^2 \mathbf{E} = j\omega\mu\sigma\mathbf{E} \tag{7.7}$$

Equations 7.6 and 7.7 are the equivalent forms of the skin effect equation. Since they are vector equations, they must hold for each term of $\mathbf{J}$ and $\mathbf{E}$.

Consider a pair of infinitely wide parallel slab conductors, oriented as shown in Figure 47. All current is assumed to be flowing in the z direction. The electric field in the conductors must also be in the z direction. The functional dependence of the electric field, however, can only be in the y direction. Equation 7.7, in this system, is

$$\frac{d^2 E_z(y)}{dy^2} = T^2 E_z(y) \tag{7.8}$$

where

$$T^2 = j\omega\mu\sigma \tag{7.9}$$

or

$$T = \sqrt{j\omega\mu\sigma} = (1+j)\sqrt{\frac{\omega\mu\sigma}{2}} = \frac{(1+j)}{\delta} \tag{7.10}$$

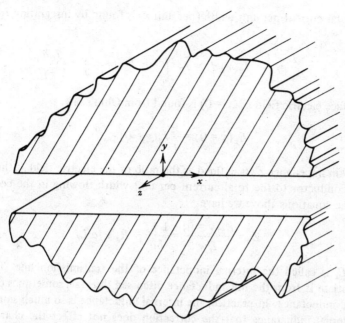

Figure 47 Parallel semi-infinite slab transmission line cross section.

Usually δ is defined as the skin depth for reasons that will become apparent shortly,

$$\delta = \frac{1}{\sqrt{\pi f \mu \sigma}} \qquad (7.11)$$

Equation 7.8, which holds for both E_z and J_z has the general solution

$$J_z(y) = C_1 e^{-Ty} + C_2 e^{+Ty} \qquad (7.12)$$

Referring to Figure 47, consider only the upper conductor. Since the current in this conductor must mirror the current in the lower conductor, there is no information lost by considering only the upper conductor. Note however that the total losses per unit length of the line must be twice the loss due to the upper conductor alone. Also, consider the upper conductor as being arbitrarily thick—that is, let it extend from $y = 0$ to $y = \infty$. The boundary conditions on (7.12) are now

$$J_z(0) = J_0 \qquad (7.13a)$$

$$J_z(\infty) = 0 \qquad (7.13b)$$

Therefore

$$J_z(y) = J_0 e^{-Ty} \qquad (7.14)$$

The total current per unit width (per unit x) is found by integrating $J_z(y)$ over all y:

$$\frac{I}{w} = \int_0^\infty J_z(y)\, dy = J_0 \int_0^\infty e^{-Ty}\, dy = \frac{J_0}{T} = \frac{J_0 \delta}{1+j} \qquad (7.15)$$

The surface electric field $E_z(y=0)$ is found from Ohm's law,

$$E_z(y=0) = \frac{1}{\sigma} J_z(y=0) = \frac{J_0}{\sigma} \qquad (7.16)$$

The skin impedance Z_s is defined as the ratio of the electric field at the surface of the conductor to the total current per unit width flowing in the conductor. From the equations above we have

$$Z_s = \frac{E_z(y=0)}{I/w} = \frac{1+j}{\delta\sigma} \equiv R_s + j\omega L_i \qquad (7.17)$$

where L_i is called the internal inductance of the transmission line. This is to differentiate it from the line inductance discussed thus far, sometimes called the external inductance. In practice, the internal inductance is so much smaller than the external inductance that the correction does not affect the characteristic impedance or the wave velocity measurably.

The symbol R_s represents the skin resistance, sometimes called the sheet resistance, of the transmission line. Equation 7.17 predicts that the skin resistance and the reactance due to the internal inductance are numerically equal. This property is of little consequence by itself, but it plays an important part in calculating R_s in many cases. Note that this equality occurs only in the case of an infinitely thick slab.

From (7.17),

$$R_s = \frac{1}{\sigma\delta} = \sqrt{\frac{\pi f \mu}{\sigma}} \qquad (7.18)$$

In (7.15) the upper limit of integration was taken to be infinity. This is in agreement with the initial assumption that the conductor was infinitely thick. An inspection of the integrand will show that the integrand exists, for all practical purposes, only for y less than approximately 5. This means that there is no current flowing more than 5 skin depths below the surface of an infinitely thick slab of conductor. Any further conclusions must be thought through very carefully, as will be shown shortly.

Equation 7.18 shows that a semi-infinite slab of material will have the same sheet resistance at a given frequency as the same material would have at dc if it were exactly one skin depth thick—hence the name skin depth. Two other observations may be made, also with reference to (7.18).

1. The value of R_s decreases only as the inverse of the square root of the conductivity of the material. This means that exhaustive efforts to get a less lossy line by increasing very slightly the conductivity of a given metallic system will probably yield disappointing results.

2. The value of R_s increases directly as the square root of frequency. This means that losses will increase with increasing frequency, and nothing can be done about it. At microwave frequencies careful attention must be paid to the preparation of the conductor surfaces—either for transmission line or wave guide. Silver plating of a highly polished surface is not unusual, nor does it contradict the foregoing statement that the slightly better conductivity of silver as compared to, say, copper is hardly worth the trouble. The issue here is that a silver surface will last, whereas a copper surface will oxidize and degrade rapidly.

A frequent misuse of (7.15) arises from the replacing of the upper limit of integration with a finite number to calculate Z_s exactly for a material of finite thickness. This makes sense only when the upper limit is so large that it does not matter exactly what the upper limit is. Physically, the same error may be caused by applying (7.18) at arbitrarily low frequencies for a finite thickness of conductor. This is equivalent to using a small upper limit of integration in (7.15), since at low enough frequencies, any finite thickness is less than 5 skin depths. Equation 7.18 predicts that any conductor will have zero resistance at dc. This is obviously an unreasonable conclusion. Only an infinitely thick slab will have zero resistance at dc. The latter conclusion does not contradict the other statements, since (7.18) was derived by assuming an infinitely thick conductor.

To find the proper skin depth relations for finite thickness slabs, it is necessary to return to the general solution of the skin effect equation (7.12), and apply the proper boundary conditions. Again, dealing only with the upper conductor of the system, assume a conductor thickness h. Since all the current in the system is flowing between $y = -h$ and $y = +h$, by Ampere's law at $y = h$ the magnetic field is zero. Since tangential magnetic fields are continuous across metallic interfaces, $H = 0$ both just inside and just outside the metal surface at $y = h$.

Returning to Maxwell's equations, expand

$$\nabla \times \mathbf{E} = -j\omega\mu\mathbf{H} \tag{7.19}$$

into its components, noting that the electric field is $E_z(y)$ only. This procedure yields

$$\frac{dE_z}{dy} = -j\omega\mu\mathbf{H}_z = 0 \tag{7.20}$$

at $y = h$. Using Ohm's law, therefore,

$$\left. \frac{dJ_z}{dy} \right|_{y=h} = 0 \tag{7.21}$$

Equation 7.21 is the proper upper boundary condition needed for (7.12). Using this and (7.13a) as boundary conditions for (7.12), we have

$$J_0 = C_1 + C_2 \tag{7.22a}$$

$$\frac{dJ(h)}{dy} = 0 = T(-C_1 e^{-Th} + C_2 e^{+Th}) \tag{7.22b}$$

This pair of equations can be solved simultaneously. Substituting the solution into (7.12) gives

$$J_z(y) = \frac{J_0 \cosh\,(T(y-h))}{\cosh\,(Th)} \tag{7.23}$$

As in the case of the infinitely thick slab, the total current per unit width is found by integrating J_z over y:

$$\frac{I}{w} = \int_0^h J_z(y)\,dy = \frac{\sigma J_0}{T} \tanh\,(Th) \tag{7.24}$$

and then

$$Z_s = \frac{E_0}{I/w} = \frac{T}{\sigma} \coth\,(Th) \tag{7.25}$$

The real and imaginary parts of Z_s are

$$R_s = \mathrm{Re}(Z_s) = \frac{1}{\sigma\delta}\left(\frac{\sinh\,(x) + \sin\,(x)}{\cosh\,(x) - \cos\,(x)}\right) \tag{7.26}$$

$$\omega L_i = \mathrm{Im}(Z_s) = \frac{1}{\sigma\delta}\left(\frac{\sinh\,(x) - \sin\,(x)}{\cosh\,(x) - \cos\,(x)}\right) \tag{7.27}$$

where $x = 2h/\delta$.

Figure 48 shows the real and imaginary parts of Z_s, normalized to $\sigma\delta$, versus the thickness of the slab, normalized to δ. As might be expected, $R_s \to \infty$ as $h \to 0$. This solution therefore does not suffer from the nonphysical conditions that the semi-infinite slab solution exhibited when used incorrectly. The internal inductance goes to zero as h goes to zero, and both the internal inductance and the sheet resistance go to $1/\sigma\delta$ for h large. For h greater than approximately 4δ both R_s and L_i essentially take on the semi-infinite slab values.

An interesting point is that (7.26) predicts that the normalized value of R_s actually dips below unity, falling to ≈ 0.92 at $h = \pi/2$ skin depths. The reason for this dip can be understood by examining the current density profile in the metal for several thicknesses of metal. Figures 49 and 50 show the magnitude and angle of $J_z(y)$, normalized to J_0, respectively. Consider a case where h is several

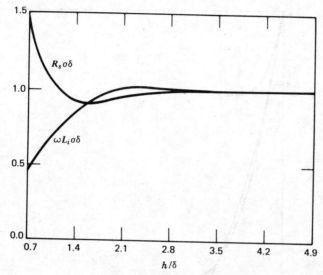

Figure 48 $R_s \sigma \delta$ and $\omega L_i \sigma \delta$ versus h/δ.

skin depths. Note that at $y = \pi/2$ skin depths the angle of J_0 passes through -90 degrees. That is, at $y = \pi/2$ skin depths, the reactive component of J_z reverses in sense. Each increment dy in the slab can be thought of as a series $dR_s + jdL_i$. The total slab consists of an infinite sum of these incremental series circuits, connected in parallel. For angles between -90 and -180 degrees the current is out of phase with the current in the incremental sections in the region from the surface to $y = \pi/2$ skin depths, where the angle is between 0 and -90 degrees. In other words, the current is flowing in the opposite direction in these two regions and the net result is to decrease the total current flow—that is, to increase the total impedance as seen from the surface.

Figure 49 shows that the magnitude of J_z falls monotonically with increasing y. This means that the current reversal does not completely cancel all current flow, but merely causes a small "swing" in the curve. Since the phase reversal is periodic, a magnified sketch of R_s versus h would show a heavily damped cyclic curve asymptotically approaching unity. The minimum value of R_s is achieved exactly once, at $\pi/2$ skin depths.

The results discussed above are exact only for the case derived—that of a pair of infinitely wide plates. These results cannot be generalized because for most practical lines, there is an x as well as a y dependence of J_z, and the skin effect partial differential equation (7.7) becomes very difficult to solve.

The closest case to the ideal situation just discussed is that of the coaxial cable. Rigorously, the coaxial cable has a circular cross section, and the skin effect equation should be solved in cylindrical coordinates. This is not a particularly

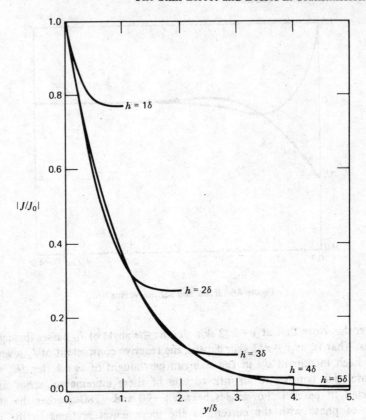

Figure 49 $|J_z(y)|/|J_0(y)|$ for several values of h.

difficult job, but it is usually unnecessary because at most frequencies where skin effect losses are of interest, the skin depth is so much smaller than the inner conductor radius that the inner conductor appears flat and infinite in extent when viewed from the scale of several skin depths. The concept "infinite in extent" is of course taken to mean that there is no lateral dependence of J_z, and the exact solution would not differ in any meaningful sense from the approximate rectangular solution.

The sheet resistance per unit length of the inner conductor of a coaxial cable, using the rectangular solution assumption, is simple $R_s/2\pi a$, where a is the inner conductor radius, and R_s is given exactly by (7.26) and approximately by (7.18) for $h \gtrsim 4\delta$. Similarly, the sheet resistance per unit length of the outer conductor is $R_s/2\pi b$, where b is the outer conductor radius. The total resistance of the cable is the sum of the two resistances above.

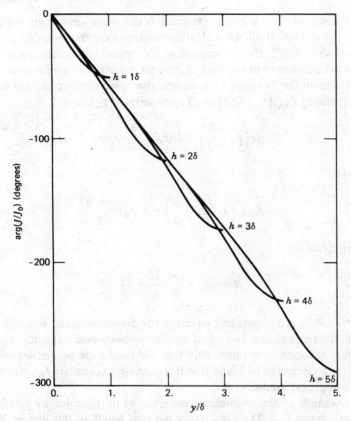

Figure 50 Angle of $J_z(y)/J_0$ for several values of h.

7.2 THE INCREMENTAL INDUCTANCE CALCULATION

Equation (7.17) showed that in the case of a thick slab of conductor, the sheet resistance and the internal inductive reactance are equal. That is,

$$R_s = \omega L_i \qquad (7.28)$$

Solving for L_i, using (7.18), we get

$$L_i = \frac{R_s}{\omega} = \frac{\mu\delta}{2} \qquad (7.29)$$

The internal inductance is caused by the penetration of a magnetic field into the conductor. As (7.29) shows, it is the inductance of a layer of conductive material $\delta/2$ thick.

If the conductors in a given transmission line were ideal, there would be no internal inductance. If all these ideal conductors were then receded by an incremental thickness $\delta/2$, the corresponding incremental inductance would be just the internal inductance of the same system built using real conductors.

Expanding on the foregoing idea, assume that some given conductor dimension is z_0. Expanding $L_{ext}(z_0 - \delta/2)$ in a Taylor series, if $z_0 \gg \delta/2$,

$$L\left(z_0 - \frac{\delta}{2}\right) \simeq L(z_0) - \frac{\partial L}{\partial z}\frac{\delta}{2} + \cdots \tag{7.30}$$

Therefore,

$$L_{int} = L\left(z_0 - \frac{\delta}{2}\right) - L(z_0) = \frac{\delta}{2}\frac{\partial L}{\partial z} \tag{7.31}$$

and using (7.28),

$$R = \omega L_i = \frac{\omega\delta}{2}\frac{\partial L}{\partial z} = \frac{R_s}{\mu}\frac{\partial L}{\partial z} \tag{7.32}$$

Equation 7.32 is a convenient equation for determining the skin resistance of an arbitrary transmission line cross section without ever explicitly solving the skin effect equation. It assumes only that the conductor geometries are large in cross section compared to δ, and that it is possible to calculate L_{ext} for the same system using ideal conductors.

As an example of the incremental inductance rule, consider the parallel round wire transmission line. The capacitance per unit length of this line, as derived in Chapter 6, is

$$C = \frac{\epsilon\pi}{\cosh^{-1}(a/b)} \tag{7.33}$$

where a = radius of each round wire
$\quad\quad 2b$ = center-to-center separation of the wires

The (external) inductance of this line is

$$L = \frac{\mu\epsilon}{C} = \frac{\mu}{\pi}\cosh^{-1}\left(\frac{b}{a}\right) \tag{7.34}$$

The required derivative is therefore

$$\frac{\partial L}{\partial a} = \frac{\mu}{\pi}\frac{\partial}{\partial a}\cosh^{-1}\left(\frac{b}{a}\right) = \frac{\mu}{\pi a}\frac{1}{\sqrt{1 - a^2/b^2}} \tag{7.35}$$

and

$$R = \frac{R_s}{\pi a} \frac{1}{\sqrt{1 - a^2/b^2}} \tag{7.36}$$

This result can be examined physically by considering two extreme cases. First, for $a \ll b$, $R \simeq R_s/\pi a$. This means that when the wires are very far apart the actual separation simply does not matter. Also, R in this case is exactly twice the resistance found for the inner (or the outer) conductor of the coaxial cable in the previous section. Since two wires are being considered in this case, this result is reasonable. Furthermore, when the two wires are very far apart, in each case the magnetic fields surrounding one wire have very little effect on the other wire. Here the current distribution in each wire should be independent of the location of the other wire, and symmetry conditions then dictate that the current be uniformly distributed about the circumference of each wire. This distribution is identical to that of the coaxial cable.

When the two wires are close together (7.36) shows that the resistance goes up quickly with decreasing b. This is because bringing the two wires near each other causes the currents to bunch on the portions of the wires nearest to each other. Since the currents are now flowing through smaller areas than previously, the resulting resistance is naturally higher. This bunching of currents is due to the proximity effect.

In the case of stripline and microstrip transmission lines, the resistance is dominated by bunching of the current that occurs at the edges of the center (or upper in the case of microstrip) conductor. The calculations for finding R in these cases is straightforward, although somewhat lengthy. The references at the end of this chapter contain both calculations. Chapter 10 gives a more detailed description of bunching currents in stripline and microstrip.

7.3 SUGGESTED FURTHER READING

1. S. Ramo et al., *Fields and Waves in Communications Electronics*, 2nd ed., Wiley, New York, 1965. Contains several solutions to circular and multilayer metal problems. The second edition (1959) contains some detail that was dropped in the 1965 edition.

2. E. Jordan and K. Balmain, *Electromagnetic Waves and Radiating Systems*, Prentice-Hall, Englewood, Cliffs, N.J., 1968. Verifies the validity of the skin effect equation, and solves the exact problem of a propagating wave with real metal boundary conditions.

3. H. Wheeler, "Formulas for the Skin Effect," *Proceedings of the IRE*, Vol. 30, No. 9, September 1942. The incremental inductance rule was first presented in this paper.

4. H. Howe, Jr., *Stripline Circuit Design*, Artech House, Dedham, Mass., 1974. Gives formulas and extensive charts for stripline losses. Also considers various typical dielectric materials.

5. R. Pucel et al., "Losses in Microstrip," *IEEE Transactions on Microwave Theory and Technique*, Vol. MTT-16, No. 6, June 1968. Applies the incremental inductance rule to the quasi-static microstrip formulas. Also see the correction, Vol. MTT-16, No. 12, December 1968.

8

Coupled Transmission Lines

When two lengths of transmission line are placed in close enough proximity for their fields to interact, the lines are said to be coupled. The properties of a pair of coupled lines have no direct analogy in lumped element network theory. Since modes of propagation are being coupled, there is a directivity as well as a magnitude and a phase associated with the couplings. This phenomenon leads to four-port networks known as directional couplers, which were introduced briefly in Chapter 4.

A pair of coupled lines, in general, forms a four-port network. By grounding or opening ("floating") various combinations of two ports, the transfer characteristics between the remaining two ports may be made to show one of several basic two-pole filter characteristics—if the lines are lightly coupled. For heavily coupled lines the filter characteristics can be classified into types but do not have simple correspondences in lumped circuit theory. Both the lightly and the heavily coupled versions of these filters are usually easier to realize at UHF and at microwave frequencies than their closest lumped element counterparts.

8.1 THE FOUR–PORT IMPEDANCE MATRIX
FOR A COUPLED PAIR OF IDENTICAL LINES

Consider a pair of identical transmission lines, equal in length, running parallel for a distance h in a uniform homogeneous medium. Figure 51 is a schematic representation of these lines (for convenience, the ground returns are not shown). As a network, this pair of lines forms a symmetric four-port. Because of the symmetries involved, there can be no more than two parameters necessary to characterize this four-port network. These parameters will of course be functions of the propagation constant of the lines, the frequency, and the length of the lines, as well as the distance between them and the geometries.

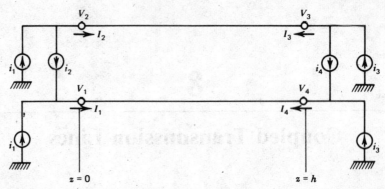

Figure 51 Voltage and current conventions for a pair of coupled lines. Ground return line *not* shown.

A convenient, but not necessary, choice of parameters is two that will converge to Z_0, the characteristic impedance of one of the lines, when the lines are separated. Consider driving a pair of lines extending to $z = \infty$ with currents that are equal in magnitude and either equal or opposite in sign. In both cases the same impedance is seen looking into either line (measured between that line and ground). When the signs of the current sources are the same, define the impedance measured as the *even mode* characteristic impedance, Z_{0e}. When the signs of the current sources are opposite, define the impedance measured as the *odd mode* characteristic impedance, Z_{0o}. Since any arbitrary pair of driving sources can always be expressed as a linear sum of the two combinations of driving sources above, it follows that any impedance measured (therefore any transfer function) must be expressable as a linear combination of Z_{0e} and Z_{0o}.

Digressing for a moment, consider the general relation for the voltage along an open-circuited line of length h driven by a current source at $z = 0$. Since $I(h) = 0$, from (3.2),

$$A_1 e^{-\gamma h} = A_2 e^{\gamma h} \tag{8.1}$$

Substituting this into (3.1),

$$V(z) = A_1 [e^{-\gamma z} + e^{-2\gamma h} e^{\gamma z}] \tag{8.2}$$

Returning to (8.1) and solving for A_1,

$$A_1 = \frac{Z_0 I_0}{1 - e^{-2\gamma h}} \tag{8.3}$$

The voltage along the line (assuming $\alpha = 0$) is therefore

$$V(z) = -j Z_0 I_0 \frac{\cos [\beta(h - z)]}{\sin (\beta h)} \tag{8.4}$$

From (8.4) we know that the voltages at either end of the line are

$$V(0) = -jZ_0I_0 \cot(\beta h) \tag{8.5}$$

$$V(h) = -jZ_0I_0 \csc(\beta h) \tag{8.6}$$

Similarly, for a line open-circuited at $z = 0$ and driven by a current source I_0 at $z = h$,

$$V(0) = -jZ_0I_0 \csc(\beta h) \tag{8.7}$$

$$V(h) = -jZ_0I_0 \cot(\beta h) \tag{8.8}$$

Returning to the four-port network problem, by superposition the voltage response due to any number of current sources connected at either end can always be written as the sum of the responses to the individual current sources. These responses are one or more of the (four) equations above. Figure 51 shows a particular case of this general situation. Sources i_1 and i_3 (two of each) both excite a symmetric voltage wave form on both lines. This is the even mode excitation, therefore the voltages due to i_1, denoted $V_{1,1}$, $V_{2,1}$, and so on, and i_3 are as follows:

$$V_{1,1} = V_{2,1} = -ji_1Z_{0e} \cot(\beta h) \tag{8.9a}$$

$$V_{3,1} = V_{4,1} = -ji_1Z_{0e} \csc(\beta h) \tag{8.9b}$$

$$V_{1,3} = V_{2,3} = -ji_3Z_{0e} \csc(\beta h) \tag{8.9c}$$

$$V_{3,3} = V_{4,3} = -ji_3Z_{0e} \cot(\beta h) \tag{8.9d}$$

Current sources i_2 and i_4 each excite antisymmetric voltage wave forms on the lines. This is the odd mode, and the terminal voltages due to i_2 and i_4 are

$$V_{1,2} = -V_{2,2} = -ji_2Z_{0o} \cot(\beta h) \tag{8.10a}$$

$$V_{4,3} = -V_{3,3} = -ji_4Z_{0o} \cot(\beta h) \tag{8.10b}$$

$$V_{1,4} = -V_{2,4} = -ji_4Z_{0o} \csc(\beta h) \tag{8.10c}$$

$$V_{4,2} = -V_{3,2} = -ji_2Z_{0o} \csc(\beta h) \tag{8.10d}$$

The total voltage at each node is simply the sum of all the contributions. That is,

$$V_i = \sum_{j=1}^{4} V_{i,j} \quad \text{for} \quad 1 \leqslant i \leqslant 4 \tag{8.11}$$

The terminal currents are, by inspection,

$$I_1 = i_1 + i_2 \tag{8.12a}$$

$$I_2 = i_1 - i_2 \tag{8.12b}$$

$$I_3 = i_3 - i_4 \tag{8.12c}$$

$$I_4 = i_3 + i_4 \tag{8.12d}$$

To arrive at the Z matrix for the four-port network, it is necessary to derive four equations relating the terminal voltages to the terminal currents. Equation 8.11 relates the terminal voltages to the source currents, and equations (8.12) relate the terminal currents to the source currents. Inverting (8.12) gives

$$i_1 = \frac{I_1 + I_2}{2} \tag{8.13a}$$

$$i_2 = \frac{I_1 - I_2}{2} \tag{8.13b}$$

$$i_3 = \frac{I_4 + I_3}{2} \tag{8.13c}$$

$$i_4 = \frac{I_4 - I_3}{2} \tag{8.13d}$$

Substituting the four equations above as needed into the voltage expressions yields the desired results. Using matrix notation,

$$[V] = [Z][I] \tag{8.14}$$

where

$$Z_{11} = Z_{22} = Z_{33} = Z_{44} = -j\frac{(Z_{0e} + Z_{0o})}{2}\cot(\beta h) \tag{8.15a}$$

$$Z_{12} = Z_{21} = Z_{34} = Z_{43} = -j\frac{(Z_{0e} - Z_{0o})}{2}\cot(\beta h) \tag{8.15b}$$

$$Z_{13} = Z_{31} = Z_{24} = Z_{42} = -j\frac{(Z_{0e} - Z_{0o})}{2}\csc(\beta h) \tag{8.15c}$$

$$Z_{14} = Z_{41} = Z_{23} = Z_{32} = -j\frac{(Z_{0e} + Z_{0o})}{2}\csc(\beta h) \tag{8.15d}$$

This Z matrix of course can be transformed to a four-port Y matrix, S matrix, and so on.

As with any TEM system, the characteristic impedance(s) can be determined from the dc capacitances or inductances and a knowledge of the dielectric material. In the case of coupled lines, the two capacitances C_{0e} and C_{0o}, corresponding to even and odd mode excitations, are determined in principle by driving the lines with identical and then opposite source currents. In practice it is often more convenient to measure the capacitance of both lines to ground

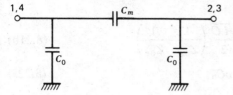

Figure 52 Dc equivalent capacitance network for a pair of symmetric lines.

and the mutual capacitance between the lines. Figure 52 shows the assumed equivalent network for this measurement. At dc, of course, these capacitance measurements completely determine the system; then Z_{0e} and Z_{0o} can be found from them.

At dc ports 2 and 3 are identical, as are ports 1 and 4. Now, as $\omega \to 0$,

$$\csc\left(\beta h\right) \approx \cot\left(\beta h\right) \simeq \frac{1}{\beta} = \frac{1}{\omega\sqrt{LC}} \tag{8.16}$$

Equations 8.15 then simplify to

$$Z_{11} = Z_{22} = Z_{33} = Z_{44} = Z_{14} = Z_{41} = Z_{23} = Z_{32} = -j\,\frac{Z_{0o} + Z_{0e}}{2\omega\sqrt{LC}} \tag{8.17a}$$

$$Z_{12} = Z_{21} = Z_{34} = Z_{43} = Z_{13} = Z_{31} = Z_{24} = Z_{42} = -j\,\frac{Z_{0o} - Z_{0e}}{2\omega\sqrt{LC}} \tag{8.17b}$$

Since at dc the network is actually two port, as was explained, there are only two independent parameters.

Consider the two lines, driven from ports 1 and 2, setting $I_3 = I_4 = 0$ in (8.14), we write

$$V_1 = Z_{11}I_1 + Z_{12}I_2 \tag{8.18a}$$

$$V_2 = Z_{12}I_1 + Z_{11}I_2 \tag{8.18b}$$

Figure 52 is a π-network model, most easily described in Y parameters. Therefore inverting (8.18) and using the approximations (8.17), we have

$$Y_{11} = j\omega\sqrt{LC}\,\frac{1/Z_{0o} + 1/Z_{0e}}{2} \tag{8.19}$$

$$Y_{12} = -j\omega\sqrt{LC}\,\frac{1/Z_{0o} - 1/Z_{0e}}{2} \tag{8.20}$$

The desired capacitances can be calculated directly from these equations (see Table 1 for the relations between the Y parameters and the π model):

$$Y_{12} = -j\omega C_m \tag{8.21a}$$

or,

$$C_m = \frac{\sqrt{LC}}{2}\left(\frac{1}{Z_{0o}} - \frac{1}{Z_{0e}}\right) \tag{8.21b}$$

$$Y_{11} + Y_{12} = j\omega C_0 \tag{8.22a}$$

and then

$$C_0 = \frac{\sqrt{LC}}{Z_{0e}} \tag{8.22b}$$

As the two lines are moved far apart,

$$Z_{0e} \simeq Z_{0o} \simeq \sqrt{\frac{L}{C}} = Z_0 \tag{8.23}$$

Using this as a check in (8.21b) and (8.22b), we have

$$C_m = 0 \tag{8.24}$$

$$C_0 = \frac{\sqrt{LC}}{\sqrt{\frac{L}{C}}} = C \tag{8.25}$$

These results are of course perfectly reasonable.

Figure 52 was introduced because it is easier to measure C_m and C_0 than C_{0e} and C_{0o} in most cases. The necessary equations to calculate C_{0e} and C_{0o} are then the inverses of (8.24) and (8.25),

$$Z_{0e} = \frac{\sqrt{LC}}{C_0} \tag{8.26}$$

$$Z_{0o} = \sqrt{LC}\left(\frac{1}{2C_m} + \frac{1}{C_0}\right) \tag{8.27}$$

As a reminder, note that $LC = \mu\epsilon = 1/v^2$.

8.2 THE BIDIRECTIONAL COUPLER

Consider a pair of identical coupled lines connected as in Figure 53. All four ports are terminated in an impedance Z_0, and port 1 is driven by a voltage source. The currents I_1 through I_4 can be found directly from the four-port Z matrix for the coupled lines. As a compromise between showing directional coupler phenomena with a reasonable amount of accuracy and deriving the general transfer equations with an unwieldy amount of algebra, consider finding

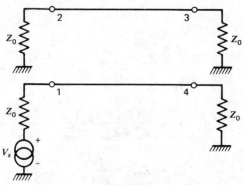

Figure 53 A bidirectional coupler.

the four terminal currents only at a single frequency. It is asserted without proof here that the frequency chosen is the center frequency for a range of operation.

The loop equations for the circuit of Figure 53 are as follows:

$$
\begin{bmatrix} V_1 \\ V_2 \\ V_3 \\ V_4 \end{bmatrix} = \begin{bmatrix} 1 \\ 0 \\ 0 \\ 0 \end{bmatrix} = \begin{bmatrix} Z_{11} + Z_0 & Z_{12} & Z_{13} & Z_{14} \\ Z_{12} & Z_{11} + Z_0 & Z_{14} & Z_{13} \\ Z_{13} & Z_{12} & Z_{11} + Z_0 & Z_{14} \\ Z_{14} & Z_{13} & Z_{12} & Z_{11} + Z_0 \end{bmatrix} \begin{bmatrix} I_1 \\ I_2 \\ I_3 \\ I_4 \end{bmatrix}
$$

$$(8.28)$$

Letting $\beta h = \pi/2$,

$$\det (Z) = Z_0^2 (Z_0^2 + Z_{0e}^2 + Z_{0o}^2) + Z_{0e}^2 Z_{0o}^2 \qquad (8.29)$$

and

$$I_1 = \frac{Z_0 (Z_0^2 + [Z_{0e}^2 + Z_{0o}^2]/2)}{\det (Z)} \qquad (8.30)$$

Now, Z_{in} as seen by the voltage source (with internal resistance Z_0) is $1/I_1 - Z_0$. For a proper match, $Z_{in} = Z_0$. By direct substitution of (8.30) into this matching condition,

$$Z_0^2 = Z_{0o} Z_{0e} \qquad (8.31)$$

Because of the symmetry of the system, the same relation holds at all four ports.

Continuing with the solution of (8.28),

$$I_2 = \frac{Z_{0o}^2 - Z_{0e}^2}{2Z_0^3 + Z_0 (Z_{0e}^2 + Z_{0o}^2)} \qquad (8.32)$$

$$V_2 = -Z_0 I_2 = \frac{1}{2} \frac{R - 1}{R + 1} \qquad (8.33)$$

where

$$R \equiv \frac{Z_{0e}}{Z_{0o}} \tag{8.34}$$

$$I_3 = 0 \tag{8.35}$$

$$I_4 = \frac{j(Z_{0e} + Z_{0o})}{2Z_0^3 + Z_{0e}^2 + Z_{0o}^2} \tag{8.36}$$

$$V_4 = -Z_0 I_4 = \frac{-j\sqrt{R}}{R + 1} \tag{8.37}$$

The $-j$ in (8.37) represents the phase shift along the line, since at $\beta h = \pi/2$ the lines are 1 quarter-wavelength long.

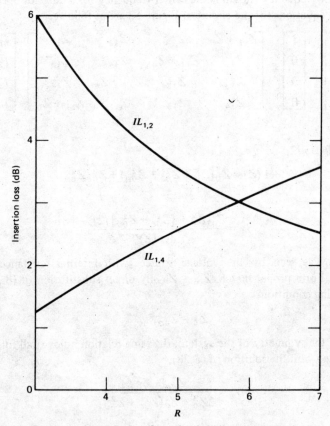

Figure 54   Forward and reverse loss for a simple bidirectional coupler versus R.

The insertion losses from ports 1 to 2 and 1 to 4 are

$$IL_{1,2} = -20 \, \text{Log}_{10} \left(\frac{R-1}{R+1} \right) \quad dB \tag{8.38}$$

$$IL_{1,4} = -20 \, \text{Log}_{10} \left(\frac{2\sqrt{R}}{R+1} \right) \quad dB \tag{8.39}$$

Since $I_3 = 0$, the insertion loss from port 1 to port 3 is infinite.

Equations 8.38 and 8.39 are plotted for a range of R in Figure 54. The symbol $IL_{1,4}$ represents energy that is lost to the direct signal path from port 1 to port 4 because of coupling to port 2; $R = 0$ corresponds to the uncoupled line case (infinitely far apart), and this loss goes to zero, assuming lossless lines. The symbol $IL_{1,2}$ represents energy that is coupled out of the signal path from port 1 to port 4, and appears at port 2. For a given value of R, port 2 "samples" the power flow from ports 1 to 4 but is totally insensitive to power flowing from port 4 to port 1. This statement may be understood easily by considering that no power appears at port 3 because of power flow from port 1 to port 4, and for the mirror image situation the same reasoning must apply.

For R less than 2, the coupling is less than approximately 10 dB, and the directional coupler is useful for sampling power flow, with little power loss to consider in the "signal" path. For R greater than 2 the coupler is also useful as a power splitter. If $R = 5.85$, for example, the power flowing into port 1 is equally divided between ports 2 and 4.

8.3 COUPLED TRANSMISSION LINE FILTERS

The basic four-port network shown in Figure 51 can be reduced to a two-port network by open-circuiting and/or grounding (short-circuiting) various combinations of ports, two at a time, and using the remaining two ports as input and output. Topologically there are 10 different combinations possible. The two-port parameter sets for these networks are found by applying the appropriate boundary conditions to the four-port Z matrix. A convenient method of discovering the frequency-response characteristics of each of these two-port networks is to calculate the image impedance, as defined in Chapter 1, for each of these networks. Figure 55 shows the 10 possible interconnections and identifies the response of each.

As an example of the actual response of one of these two-port networks, consider Case 2 of Figure 55, where $V_2 = V_4 = 0$, and ports 1 and 3 are used as input and output. Since we are eliminating voltages variables from the four-port network, it is much more convenient to work with a four-port Y matrix than the four-port Z matrix derived in Section 8.1. The Y matrix can be obtained directly

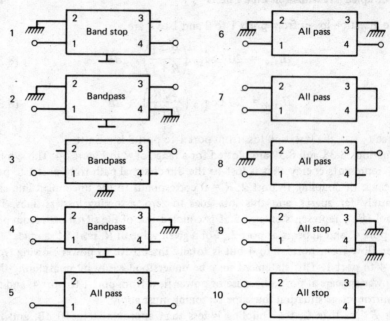

Figure 55 Ten topological interconnection possibilities for forming a two-port network from a pair of coupled lines.

from the basic circuit by assuming voltage sources and paralleling the Z-matrix derivation, or by inverting the Z matrix already derived. In either case, the result is

$$Y_{11} = Y_{22} = Y_{33} = Y_{44} = -j\frac{(Y_{0o} + Y_{0e})}{2}\cot(\beta h) \qquad (8.40a)$$

$$Y_{12} = Y_{21} = Y_{34} = Y_{43} = -j\frac{(Y_{0o} - Y_{0e})}{2}\cot(\beta h) \qquad (8.40b)$$

$$Y_{13} = Y_{31} = Y_{24} = Y_{42} = -j\frac{(Y_{0o} - Y_{0e})}{2}\csc(\beta h) \qquad (8.40c)$$

$$Y_{14} = Y_{41} = Y_{23} = Y_{32} = -j\frac{(Y_{0o} + Y_{0e})}{2}\csc(\beta h) \qquad (8.40d)$$

where Y_{0o} and Y_{0e} follow directly from the definitions of Z_{0o} and Z_{0e}.

For the example chosen, setting V_2 and V_4 equal to zero in (8.40) leaves four current equations in two unknowns. However the currents I_2 and I_4 are of no use insofar as the two-port response at ports 1 and 3 is concerned. Therefore, the only remaining equations of interest are

$$I_1 = Y_{11} V_1 + Y_{13} V_3 \tag{8.41a}$$

$$I_3 = Y_{13} V_1 + Y_{11} V_3 \tag{8.41b}$$

Since Y_{11} and Y_{33} are zero at $\beta h = \pi/2$, one would expect some sort of peak in the transfer response at that point. Intuitively, for the moment, define the frequency at which $\beta h = \pi/2$ as the "center frequency" of the network response. Considering this center frequency response in more detail, at this frequency

$$Y_{11} = 0 \tag{8.42}$$

$$Y_{13} = -j \frac{(Y_{0o} - Y_{0e})}{2} \tag{8.43}$$

From two-port network theory, for a symmetric network terminated in a conductance G_s and driven from a source with internal conductance G_s, the insertion loss (IL) is given by

$$IL = \left[\frac{2G_s Y_{13}}{(Y_{11} + G_s)^2 - Y_{13}^2} \right]^2 \tag{8.44}$$

Substituting (8.42) and (8.43) into (8.44), we have

$$IL_{center} = \left[\frac{Y_s (Y_{0o} - Y_{0e})}{Y_s^2 + (Y_{0o} - Y_{0e}^2)/4} \right]^2 \tag{8.45}$$

Assuming that the coupled transmission lines themselves are lossless (this assumption has been made throughout the chapter—there is no loss term included in the Y or Z matrix parameters), (8.45) should be set equal to unity if the network is properly matched. Saying the same thing from a slightly different point of view, setting (8.45) equal to unity determines the matching condition:

$$Y_s (Y_{0o} - Y_{0e}) = Y_s^2 + \frac{Y_{0o}^2 - Y_{0e}^2}{4} \tag{8.46a}$$

from which

$$Y_s = \frac{Y_{0o} - Y_{0e}}{2} \tag{8.46b}$$

If this two-port network is to be a filter of some sort, then (8.46b) can be considered to be the first design equation for the filter. It relates the two line parameters to the source (and terminating) impedance.

A useful parameter to define at this point is the ratio of the odd and even mode admittances, R:

$$R \equiv \frac{Y_{0e}}{Y_{0o}} = 1 + 2 \frac{Y_s}{Y_{0e}} \qquad (8.47)$$

This equation predicts that R is never less than unity, that is, that Y_{0o} is never less than Y_{0e}. This is due to the basic physical properties of the coupled line system and has nothing whatsoever to do with this particular example. A proof is given below.

Figure 56 shows the insertion loss, as calculated using (8.44), (8.46), and (8.47), versus frequency, for several values of R. Clearly, the coupled line pair as connected in this example behaves as a bandpass filter. The parameter R in this case is a measure of the bandwidth. An important difference between coupled line filters of this sort and lumped element filters arises because the matrix

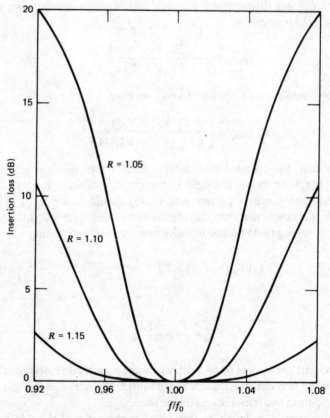

Figure 56 Insertion loss of simple two-pole coupled line filter versus frequency for several values of R.

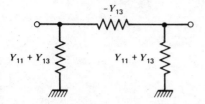

Figure 57 A π model of a coupled line pair.

parameters—in this case Y_{ij}—are periodic in f. A coupled line bandpass filter therefore has an infinite number of passbands. Figure 56 shows only the lowest passband frequency region.

To compare the coupled line bandpass filter to a lumped element filter, it would be helpful to be able to construct a lumped element model for the coupled line filter. In general this cannot be done. However for a narrow frequency range about center frequency, a reasonably good approximation is possible.

Since the filter was described in Y parameters, it is easiest to deal with admittance parameters throughout and to model the filter near center frequency using a π model, as in Figure 57. Near center frequency, that is, for $\beta h \approx \pi/2$,

$$-Y_{13} = j\frac{(Y_{0o} - Y_{0e})}{2} \csc(\beta h) \simeq j\frac{(Y_{0o} - Y_{0e})}{2} \qquad (8.48)$$

and

$$Y_{11} + Y_{13} = -j\left[\frac{(Y_{0o} + Y_{0e})}{2}\cot(\beta h) + \frac{(Y_{0o} - Y_{0e})}{2}\csc(\beta h)\right] \qquad (8.49)$$

$$= -\frac{j}{2}\left[Y_{0o}(\cot(\beta h) + \csc(\beta h)) + Y_{0e}(\cot(\beta h) - \csc(\beta h))\right]$$

Now we can write

$$\cot(\beta h) \pm \csc(\beta h) = \frac{\cos(\beta h) \pm 1}{\sin(\beta h)} \qquad (8.50)$$

For a lossless line, $\beta h = \omega\sqrt{LC}h$, where $\omega_0\sqrt{LC}h = \pi/2$. For $\omega \approx \omega_0$,

$$\cot(\beta h) \pm \csc(\beta h) \simeq \sqrt{LC}h(\omega_0 - \omega) \pm 1 \qquad (8.51)$$

Therefore,

$$Y_{11} + Y_{13} = j\left[\frac{\pi(Y_{0o} + Y_{0e})}{4\omega_0}(\omega - \omega_0) - \frac{Y_{0o} - Y_{0e}}{2}\right] \qquad (8.52)$$

Consider a two-pole, lumped element bandpass filter, capacitively coupled (Figure 58). If the uncoupled resonant frequency of each LC tank circuit is

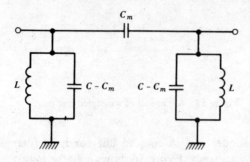

Figure 58 An equivalent lumped element filter to the simple two-pole coupled line filter.

$\omega_0 = 1/\sqrt{LC}$, then

$$Y_{11} = j\left(\omega C - \frac{1}{\omega L}\right) \qquad (8.52a)$$

$$Y_{12} = -j\omega C_m \qquad (8.52b)$$

Equating (8.52b) and (8.48),

$$C_m = \frac{Y_{0o} - Y_{0e}}{2\omega_0} \qquad (8.53)$$

Equation 8.52a does not obviously compare with (8.52). However near ω_0 we can make the approximation

$$\omega C - \frac{1}{\omega L} = \frac{\omega C(\omega - \omega_0)(\omega + \omega_0)}{\omega^2} \simeq \frac{2\omega^2 C(\omega - \omega_0)}{\omega^2} = 2C(\omega - \omega_0) \quad (8.54)$$

Now, making the desired comparison,

$$C = \frac{\pi(Y_{0o} + Y_{0e})}{8\omega_0} \qquad (8.55)$$

and, of course,

$$L = \frac{1}{\omega_0^2 C} \qquad (8.56)$$

Equations 8.46b, 8.53, 8.55, and 8.56 form a complete set of equations that relate the lumped element model to the actual circuit. As an example of the comparison, let

$$G_s = 1, \quad R = 1.10, \quad Y_{0e} = 20, \quad \omega_0 = 2\pi$$

Calculating the necessary parameters,

$$Y_{0o} = 22, \quad C_m = .159, \quad C = 2.63, \quad L = 9.65 \times 10^{-3}$$

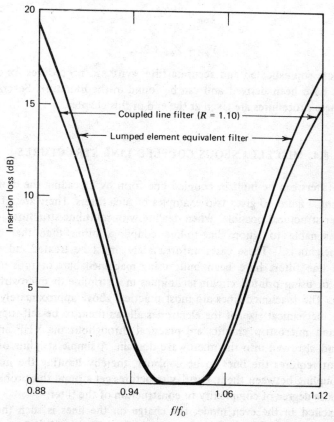

Figure 59 Insertion loss of $R = 1.1$ coupled line filter versus frequency as compared to the lumped element equivalent model.

Figure 59 shows the insertion loss for the actual $R = 1.1$ coupled line filter, and the insertion loss for the "equivalent" lumped element filter. Clearly the filters are essentially identical for insertion losses up to ≈ 15 dB, and equivalent for most practical cases for much higher insertion losses also. The filter response is that of a 12% bandwidth (approximately) two-pole bandpass filter.

To make the equivalent model just described more useful, it is necessary to invert the foregoing equations so that the coupled line filter can be found in terms of the lumped element "prototype." This is a straightforward procedure. The center frequency of the filter determines the length of the lines (quarter-wavelength), and the other three relations are given, without showing the algebra, as follows:

$$C_m = \frac{G_s}{\omega_0} \tag{8.57}$$

$$Y_{0e} = \frac{4\omega_0 c}{\pi} \tag{8.58}$$

$$Y_{0o} = Y_{0e} + 2G_s \tag{8.59}$$

Much more sophisticated and accurate filter synthesis procedures for coupled line filters have been derived and can be found in the literature. Several references to these procedures are given at the end of this chapter.

8.4 MISCELLANEOUS COUPLED LINE STRUCTURES

Multipole filters can be built in coupled line form by cascading the basic band-pass sections. Figure 60 gives two examples of such filters. There are, of course, many other structures possible. When dealing with multiline structures, it is not always reasonable to ignore line-to-line couplings other than that between "nearest neighbors." These cases unfortunately, must be treated individually.

Coupled line filters have been built using machined bars or rods in an air dielectric, or using printed circuit techniques in a stripline or microstrip form, and so on. The machined lines are most practical above approximately 3 GHz where the mechanical size of the electrodes allows them to be self-supporting. Stripline and microstrip circuits are practical throughout the VHF and UHF regions, and also well into the microwave domain. A simple stripline or microstrip layout requires the lines to be coplanar, thereby limiting the maximum level of coupling between them. Overlay structures get around this problem, but add one more degree of complexity to construction of the filter.

When excited in the even mode, the charge on the lines is such that most current flows along the outer edges of the two lines (Figure 61a). Consequently the capacitance to ground from either line is less than it would be if the line

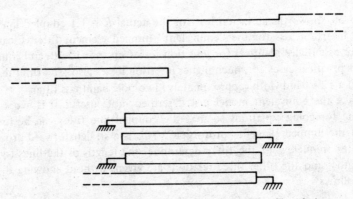

Figure 60 Two examples of multisection coupled line filter design.

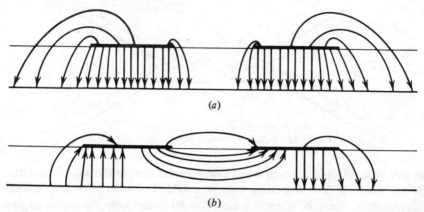

(a)

(b)

Figure 61 Symmetric (a) and asymmetric (b) electric field patterns about a pair of lines.

were "alone in the universe" with its ground return. In this case Z_{0e} must be greater than Z_0, or equivalently, Y_{0e} must be less than Y_0.

On the other hand, when the lines are excited in the odd mode, the charge on the lines concentrates on the inner edges (Fig. 61b). The capacitance of either line to ground is augmented by the capacitance to the other line, and is therefore greater than the single line capacitance to ground. In this case Z_{0o} is less than Z_0, and Y_{0o} is greater than Y_0. The assertion made in Section 8.3 that Y_{0o} must always be equal to or greater than Y_{0e} follows directly.

The non-TEM nature of microstrip results in an interesting phenomenon because of the redistribution of charge during even and odd mode excitations. In the even mode, although the capacitance to ground is less than for the single uncoupled line, since the existing electric field lines are more heavily concentrated in the dielectric than in the single line case, ϵ_{eff} is higher than for the single line; consequently the wave velocity is lower. On the other hand, in the odd mode case, some percentage of the capacitance is due to line-to-line coupling, with flux lines in the air that were not present in the single line case. Therefore ϵ_{eff} is lower, and the wave velocity is higher.

The existence of unequal wave velocities of the two modes in microstrip generally causes degradation of device parameters. Bandpass filters often perform disturbingly unlike their design prototypes—which probably did not properly treat the inhomogeneous situation. Directional couplers tend to show a lower lever of directivity than anticipated.

Figure 62 illustrates one empirical "fix" to the problem just outlined. The inner edges of the lines are zigzagged somewhat, forcing currents to redistribute themselves along a longer path length than in the simple coupled line situation. The zigzag therefore tends to lower the effective wave velocity (by lengthening the current path between the input and the output) in the odd mode case, but

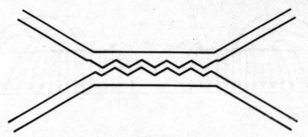

Figure 62 Zigzag microstrip directional coupler.

has very little effect in the even mode case. Coupler directivities and bandwidths, and the shape of filter responses, may be notably improved by this technique. Unfortunately, there is often a tradeoff to be made with the raising of line losses that must occur whenever current paths are lengthened.

The analysis of directional couplers presented in Section 8.2 was limited to a single frequency response. As may be expected, the performance of a simple coupler will fall off with frequency as compared with the center frequency response. Improved bandwidth can be obtained by "tandeming" couplers designed for different center frequencies. The design of a broadband coupler must take into account the interacting responses of each of the elementary couplers involved. Various design procedures have been developed for tailoring cascaded coupler section responses to different broadband characteristics. These specialized designs are beyond the scope of this treatment. References to them appear in the readings listed at the end of Chapters 6 and 8.

8.5 ASYMMETRIC COUPLED LINES

When two transmission lines that are not identical are coupled, the four-port parameter matrix describing the terminal characteristics of the lines becomes asymmetric also. The notion of even and odd impedances must be generalized to that of the two normal modes of propagation in the system. To develop the four-port Z or Y matrix for a uniformly coupled pair of uniform but not-identical lines, it is necessary to carefully examine the concepts of propagating modes on coupled lines.

Figure 63 shows a pair of uniform transmission lines, as represented by incremental circuit parameters, uniformly coupled. Following the procedure of Chapter 1, the circuit equations describing this network are as follows:

$$-\left(\frac{I_1(z + \Delta z) - I_1(z)}{\Delta z}\right) = j\omega(C_m + C'_1)V_1(z) - j\omega C_m V_2(z) \qquad (8.60)$$

$$-\left(\frac{I_2(z + \Delta z) - I_2(z)}{\Delta z}\right) = j\omega(C_m + C'_2)V_2(z) - j\omega C_m V_1(z) \qquad (8.61)$$

$$-\left(\frac{V_1(z+\Delta z)-V_1(z)}{\Delta z}\right)=j\omega L_1 I_1(z)+j\omega L_m I_2(z) \tag{8.62}$$

$$-\left(\frac{V_2(z+\Delta z)-V_2(z)}{\Delta z}\right)=j\omega L_2 I_2(z)+j\omega L_m I_1(z) \tag{8.63}$$

Taking the limit as Δz goes to zero, and writing the results in matrix form,

$$\frac{d\mathbf{I}}{dz}=-j\omega\begin{bmatrix}(C_m+C_1') & -C_m \\ -C_m & (C_m+C_2')\end{bmatrix}\mathbf{V}\equiv -j\omega C\mathbf{V} \tag{8.64}$$

$$\frac{d\mathbf{V}}{dz}=-j\omega\begin{bmatrix}L_1 & L_m \\ L_m & L_2\end{bmatrix}\mathbf{I}\equiv -j\omega L\mathbf{I} \tag{8.65}$$

where

$$\begin{bmatrix}C_1 & -C_m \\ -C_m & C_2\end{bmatrix}=C=\begin{bmatrix}C_m+C_1' & -C_m \\ -C_m & C_m+C_2'\end{bmatrix}$$

$$C_1\equiv C_m+C_1'$$

$$C_2\equiv C_m+C_2'$$

The C and L matrices can be determined from static measurements, as was the case with symmetric lines. The capacitance measurements are made using the circuits shown in Figure 64, to evaluate the terms of the C matrix using the π model as shown in Figure 65a. The three measured capacitances are related to

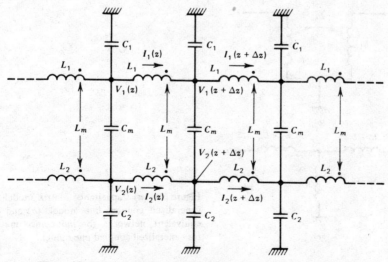

Figure 63 Incremental circuit model for asymmetric coupled lines.

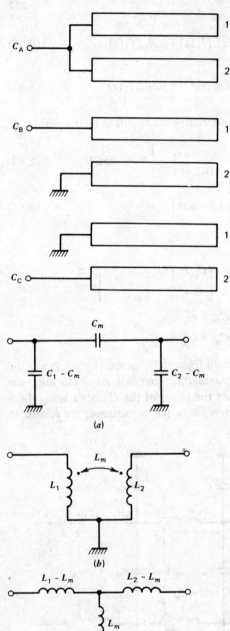

Figure 64 Three capacitance measurements for defining parameters of a pair of asymmetric coupled lines.

C_m

$C_1 - C_m$ $C_2 - C_m$

(a)

L_m

L_1 L_2

(b)

$L_1 - L_m$ $L_2 - L_m$

L_m

(c)

Figure 65 (a) Capacitance matrix model for generalized coupled line model. (b) and (c) equivalent networks for inductance matrix for generalized coupled line model.

the three desired capacitances by

$$C_A = C_1 + C_2 - 2C_m \tag{8.66a}$$

$$C_B = C_1 \tag{8.66b}$$

$$C_C = C_2 \tag{8.66c}$$

Equation 8.66a can be put into the desired (inverted) form by using (8.66a) and (8.66b):

$$C_m = \frac{C_B + C_C - C_A}{2} \tag{8.67}$$

The parameters of the inductance matrix (Figure 65b or c) can be found from the parameters of the capacitance matrix in a manner similar to that for single lines. That is, for homogeneous lines,

$$L_1 = \frac{1}{v^2 C_1} \tag{8.68a}$$

$$L_2 = \frac{1}{v^2 C_2} \tag{8.68b}$$

It is shown below that for homogeneous lines the coefficients of inductive and capacitive coupling are the same; that is,

$$k_{\text{homog}} = \frac{L_m}{\sqrt{L_1 L_2}} = \frac{C_m}{\sqrt{C_1 C_2}} \tag{8.69}$$

and L_m can be found from (8.69) in conjunction with the several equations preceding it.

In the case of inhomogeneous lines, the capacitances are found as described previously, but the inductances must be found by creating a homogeneous line with the same geometry. That is, replace all dielectrics present with air, then repeat the calculations of (8.66) through (8.69). Note, of course, that the capacitances found at this time are merely intermediate values, and the inductive coefficient of coupling will not be equal to the capacitive coefficient of coupling.

Returning to (8.64) and (8.65), the wave equation for the voltage is found by differentiating (8.65) with respect to z, and then substituting (8.64),

$$\frac{d^2 \mathbf{V}}{dz^2} + \omega^2 \, L \, C \, \mathbf{V} = 0 \tag{8.70}$$

Assuming wave solutions of the form $e^{\gamma z}$, the equation above becomes

$$\left(a + \frac{\gamma^2}{\omega^2} \right) \mathbf{V} = 0 \tag{8.71}$$

where

$$a = L C = \begin{bmatrix} (L_1 C_1 - L_m C_m) & (L_m C_2 - L_1 C_m) \\ (L_m C_1 - L_2 C_m) & (L_2 C_2 - L_m C_m) \end{bmatrix} \tag{8.72}$$

Equation 8.71 is an eigenvalue equation, and the natural modes of the system are found by setting the determinant

$$\begin{vmatrix} \left(a_{11} + \dfrac{\gamma^2}{\omega^2}\right) & a_{12} \\ a_{21} & \left(a_{22} + \dfrac{\gamma^2}{\omega^2}\right) \end{vmatrix}$$

equal to zero. This leads to solutions for γ of

$$\gamma = \pm j\omega \left[\frac{a_{11} + a_{22} \pm \sqrt{(a_{11} - a_{22})^2 + 4a_{12}a_{21}}}{2} \right]^{1/2} \tag{8.73}$$

The wave velocities of the two modes of propagation are found from (8.73), since $v = j\omega/\gamma$,

$$v = \pm \left[\frac{2}{a_{11} + a_{22} \pm \sqrt{(a_{11} - a_{22})^2 + 4a_{12}a_{21}}} \right]^{1/2} \tag{8.74}$$

In the cases of a homogeneous system, the two velocities must be equal to each other and to the wave propagation velocity in the (homogeneous) dielectric medium. Referring to (8.74), this means that

$$-(L_1 C_1 - L_2 C_2)^2 = 4(L_m C_1 - L_2 C_m)(L_m C_2 - L_1 C_m) \tag{8.75}$$

Solving for L_m, and using the fact that $L_1 C_1 = L_2 C_2 = 1/v^2$,

$$L_m = \frac{C_m}{v^2 C_1 C_2} \tag{8.76}$$

Writing L_m in terms of the coefficient of capacitive coupling, k_c,

$$L_m = \frac{C_m}{\sqrt{C_1 C_2}} \frac{1}{v^2 \sqrt{C_1 C_2}} = k_c \sqrt{L_1 L_2} \tag{8.77}$$

This equation shows that as was asserted earlier in this section, in a homogeneous medium the capacitive and inductive coefficients of coupling are equal.

To find the relations between V_1 and V_2 that, when satisfied, will drive the system exclusively at one of the two normal modes, let

$$V_1 = e^{\gamma z} \tag{8.78}$$

$$V_2 = R e^{\gamma z} \tag{8.79}$$

Substituting the relations above into (8.70), either of the resulting equations yields

$$R = -\frac{(\gamma^2 + \omega^2 a_{11})}{\omega^2 a_{12}} = \frac{1}{2a_{12}} [a_{22} - a_{11} \pm \sqrt{(a_{11} - a_{22})^2 + 4a_{12}a_{21}}] \equiv R_e, R_0$$

$$(8.80)$$

For a symmetric system, (8.80) reduces to $R = \pm 1$. These are the even and odd modes referred to earlier in this chapter. In the general case, the terms "even" and "odd" have no correspondence to the actual physical situation, but it is convenient to use the names to refer to the two propagating normal modes of the system.

The general expressions for the voltages on the two lines are as follows:

$$V_1 = A_1 e^{-\gamma e z} + A_2 e^{+\gamma e z} + A_3 e^{-\gamma_0 z} + A_4 e^{+\gamma_0 z} \tag{8.81}$$

$$V_2 = A_1 R_e e^{-\gamma e z} + A_2 R_e e^{+\gamma e z} + A_3 R_0 e^{-\gamma_0 z} + A_4 R_0 e^{+\gamma_0 z} \tag{8.82}$$

Using (8.65), the currents are found to be

$$I_1 = A_1 Y_{e1} e^{-\gamma e z} - A_2 Y_{e1} e^{+\gamma e z} + A_3 Y_{01} e^{-\gamma_0 z} + A_4 Y_{01} e^{+\gamma_0 z} \tag{8.83}$$

$$I_2 = A_1 R_e Y_{e2} e^{-\gamma e z} - A_2 R_e Y_{e2} e^{+\gamma e z} + A_3 R_0 Y_{02} e^{-\gamma_0 z} + A_4 R_0 Y_{02} e^{+\gamma_0 z}$$

$$(8.84)$$

where

$$Y_{e1} = \frac{\gamma_e(L_2 - L_m R_e)}{j\omega(L_1 L_2 - L_m^2)} = \frac{1}{Z_{e1}} \tag{8.85}$$

$$Y_{e2} = \frac{\gamma_e(L_1 R_e - L_m)}{R_e j\omega(L_1 L_2 - L_m^2)} = \frac{1}{Z_{e2}} \tag{8.86}$$

$$Y_{01} = \frac{\gamma_0(L_2 - L_m R_0)}{j\omega(L_1 L_2 - L_m^2)} = \frac{1}{Z_{01}} \tag{8.87}$$

$$Y_{02} = \frac{\gamma_0(L_1 R_0 - L_m)}{R_0 j\omega(L_1 L_2 - L_m^2)} = \frac{1}{Z_{02}} \tag{8.88}$$

The four terminal voltages (see Figure 51) are related to the four coefficients A_i, as can be seen from (8.81) and (8.82), by

$$\begin{bmatrix} V_1 \\ V_2 \\ V_3 \\ V_4 \end{bmatrix} = \begin{bmatrix} 1 & 1 & 1 & 1 \\ R_e & R_e & R_0 & R_0 \\ R_e^{-\gamma e h} & R_e e^{\gamma e h} & R_0 e^{-\gamma_0 h} & R_0 e^{+\gamma_0 h} \\ e^{-\gamma e h} & e^{+\gamma e h} & e^{-\gamma_0 h} & e^{\gamma_0 h} \end{bmatrix} \begin{bmatrix} A_1 \\ A_2 \\ A_3 \\ A_4 \end{bmatrix} \tag{8.89}$$

Similarly, the four terminal currents are related to the coefficients by

$$
\begin{bmatrix} I_1 \\ I_2 \\ I_3 \\ I_4 \end{bmatrix} = \begin{bmatrix} Y_{e1} & -Y_{e1} & Y_{01} & -Y_{01} \\ R_e Y_{e2} & -R_e Y_{e1} & R_0 Y_{02} & -R_0 Y_{02} \\ R_e Y_{e2} e^{-\gamma_e h} & -R_e Y_{e2} e^{\gamma_e h} & R_0 Y_{02} e^{-\gamma_0 h} & -R_0 Y_{02} e^{\gamma_0 h} \\ Y_{e1} e^{-\gamma_e h} & -Y_{e1} e^{\gamma_e h} & Y_{01} e^{-\gamma_0 h} & -Y_{01} e^{\gamma_0 h} \end{bmatrix} \begin{bmatrix} A_1 \\ A_2 \\ A_3 \\ A_4 \end{bmatrix}
$$

(8.90)

The four-port Y or Z parameters are found by eliminating the A_i using (8.89) and (8.90). Depending on whether (8.89) or (8.90) is inverted, the results will be the Y or the Z parameters, respectively. These parameters are

$$
Y_{11} = Y_{44} = \frac{Y_{e1} \coth (\gamma_e h)}{1 - R_e/R_0} + \frac{Y_{01} \coth (\gamma_0 h)}{1 - R_0/R_e}
$$

(8.91a)

$$
Y_{12} = Y_{21} = Y_{34} = Y_{43} = -\frac{Y_{e1} \coth (\gamma_e h)}{R_0(1 - R_e/R_0)} - \frac{Y_{01} \coth (\gamma_0 h)}{R_e(1 - R_0/R_e)}
$$

(8.91b)

$$
Y_{13} = Y_{31} = Y_{24} = Y_{42} = \frac{Y_{e1}}{(R_0 - R_e) \sinh (\gamma_e h)} + \frac{Y_{01}}{(R_e - R_0) \sinh (\gamma_0 h)}
$$

(8.91c)

$$
Y_{14} = Y_{41} = \frac{Y_{e1}}{(1 - R_e/R_0) \sinh (\gamma_e h)} - \frac{Y_{01}}{(1 - R_0/R_e) \sinh (\gamma_0 h)}
$$

(8.91d)

$$
Y_{22} = Y_{33} = \frac{-R_e Y_{e2} \coth (\gamma_e h)}{R_0(1 - R_e/R_0)} - \frac{R_0 Y_{02} \coth (\gamma_0 h)}{R_e(1 - R_0/R_e)}
$$

(8.91e)

$$
Y_{23} = Y_{32} = \frac{R_e Y_{e2}}{R_0(1 - R_e/R_0) \sinh (\gamma_e h)} + \frac{R_0 Y_{02}}{R_e(1 - R_0/R_e) \sinh (\gamma_0 h)}
$$

(8.91f)

and

$$
Z_{11} = Z_{44} = \frac{Z_{e1} \coth (\gamma_e h)}{1 - R_e/R_0} + \frac{Z_{01} \coth (\gamma_0 h)}{1 - R_0/R_e}
$$

(8.92a)

$$
Z_{12} = Z_{21} = Z_{34} = Z_{43} = \frac{Z_{e1} R_e \coth (\gamma_e h)}{1 - R_e/R_0} + \frac{Z_{01} R_0 \coth (\gamma_0 h)}{1 - R_0/R_e}
$$

(8.92b)

$$
Z_{13} = Z_{31} = Z_{24} = Z_{42} = \frac{R_e Z_{e1} \operatorname{csch} (\gamma_e h)}{1 - R_e/R_0} + \frac{R_0 Z_{01} \operatorname{csch} (\gamma_0 h)}{1 - R_0/R_e}
$$

(8.92c)

$$Z_{14} = Z_{41} = \frac{Z_{e1} \text{ csch } (\gamma_e h)}{1 - R_e/R_0} + \frac{Z_{01} \text{ csch } (\gamma_0 h)}{1 - R_0/R_e} \tag{8.92d}$$

$$Z_{22} = Z_{33} = \frac{R_e^2 Z_{e1} \text{ coth } (\gamma_e h)}{1 - R_e/R_0} + \frac{R_0^2 Z_{01} \text{ coth } (\gamma_0 h)}{1 - R_0/R_e} \tag{8.92e}$$

$$Z_{23} = Z_{32} = \frac{R_e^2 Z_{e1} \text{ csch } (\gamma_e h)}{1 - R_e/R_0} + \frac{R_0^2 Z_{01} \text{ csch } (\gamma_0 h)}{1 - R_0/R_e} \tag{8.92f}$$

where h is, of course, the length of the coupled sections of line.

As an example of coupled asymmetric lines, consider the dual concentric coaxial cable in Figure 66. Consider the dielectric to be homogeneous. Following the format of equations (8.66), the three measured capacitances are

$$C_A = \frac{2\pi\epsilon}{Ln(2/r_2)} \tag{8.93a}$$

$$C_B = \frac{2\pi\epsilon}{Ln(r_2/1)} \tag{8.93b}$$

$$C_C = C_A + C_B \tag{8.93c}$$

where the outer line is taken as the ground reference.

Equation 8.93c is somewhat unusual—it represents a drastic departure from the coupled stripline and microstripline configurations discussed earlier in this chapter. In these cases, both the coupled lines "saw" the ground return directly.

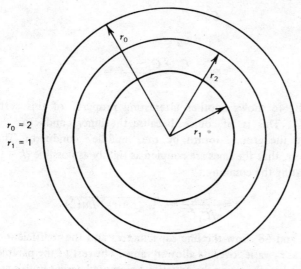

$r_0 = 2$
$r_1 = 1$

Figure 66 Dual concentric coaxial cable.

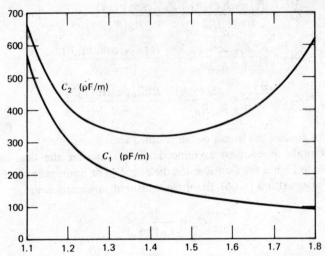

Figure 67 C_1 and C_2 versus r_2 for the dual concentric coaxial cable.

An asymmetric coupled stripline or microstripline would differ numerically but not conceptually from the symmetric case.

In the dual concentric coaxial cable, the inner conductor never "sees" ground directly. Also, as (8.93c) and (8.93b) indicate, C_B and C_C are never equal—there is no choice of r_1 that reduces this situation to that of a symmetric pair of coupled lines. The three capacitances needed for the capacitance matrix are, using (8.93) and (8.67),

$$C_B = C_1 \tag{8.94a}$$

$$C_C = C_2 \tag{8.94b}$$

$$C_m = \frac{C_1 + C_2 - C_A}{2} \equiv C_1 \tag{8.94c}$$

Equation 8.94c shows another interesting property of this system: C_m is identically C_1. This is, of course, because the inner conductor is completely isolated from the ground return by the "middle" conductor. This does not mean, however, that the lines are coupled as tightly as possible ($k = 1$) regardless of r_2. Calculating the coupling,

$$k = \frac{C_m}{\sqrt{C_1 C_2}} = \sqrt{1 - Ln(r_2)/Ln(2)} \tag{8.95}$$

Figures 67 and 68 show the line capacitances and the coefficient of coupling, respectively, as r_2 varies over its allowed range. The rest of the parameters of this system follow directly, using the equations presented earlier in this section.

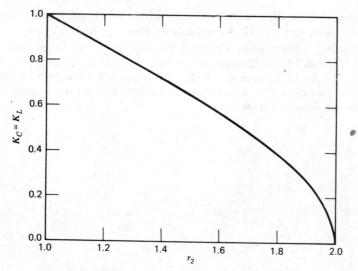

Figure 68 k versus r_2 for dual concentric coaxial cable.

The asymmetric coupled line system is useful for all the same reasons as the symmetric coupled line system, with the added advantage that the asymmetric system will also match to different impedances at the ports terminating each of the different impedance lines. One practical use of this ability is the design of a dual directional coupler that samples power flow in some arbitrary characteristic impedance system and provides its samples at a characteristic impedance common to the measuring system available. Another use is based on the fact that in many cases, materials considerations show that a filter may be built less lossy at a much lower impedance than what would be required if the filter line sections were at the needed impedance. In this case an N section filter is designed, as needed, at the lowest impedance that space considerations allow; then two additional lines are added, these lines being designed at the desired input and output impedance. The last two lines form broad passband filters with the first and last filter lines.

8.6 SUGGESTED FURTHER READING

1. Leo Young, Editor, *Parallel Coupled Lines and Directional Couplers*, Artech House, Dedham, Mass., 1972. This collection of reprints from the literature is an excellent basis for a detailed study of parallel coupled transmission lines. In particular, the paper by Jones and Bolljahn details all 10 coupled line topologies and their image parameters (for symmetric lines).

2. Leo Young, Editor, *Microwave Filters Using Parallel Coupled Lines*, Artech House, Dedham, Mass., 1972. A companion volume to Reference 1.

3. Vijai Tripathi, "Asymmetric Coupled Transmission Lines in an Inhomogeneous Medium," *IEEE Transactions on Microwave Theory and Techniques*, Vol. MTT-23, No. 9, September 1975. Line parameters for asymmetric lines are worked out explicitly, as are some prototype two-port interconnections of the general four-port system.

9

Discrete Variable Solutions of Laplace's Equation

Although in many cases the available analytic solutions for single and coupled transmission line parameters in terms of geometry and dielectric constant are elegant, they are unfortunately very limited. Except for several special cases it must be concluded that practical problems are unsolvable analytically. This may be aesthetically disappointing, but it does not close the door to finding solutions to the real problems that arise. Quite a few numerical techniques are applicable to transmission line problems. In some situations the algorithms are so efficient that a computer can generate results as quickly as it (the same computer) could evaluate an analytic solution, assuming one were available, to the same problem.

This chapter considers an important numerical approach to transmission line problems, that of solving Laplace's equation numerically on a grid of points. It is shown that a two-dimensional grid can be used to describe the potential function for an arbitrarily shaped uniform transmission line quite efficiently. Once the potential function is known, the capacitance(s) per unit length, then the characteristic impedance(s), are easily calculated.

In the material that follows, several proofs of convergence of numerical procedures have been omitted. These proofs can be found in the references cited at the end of the chapter. On the other hand, detailed examples are given. These examples are written in standard FORTRAN IV because at the time of writing, FORTRAN IV was the most widely used and accepted scientific computer language. In many cases obvious (and perhaps not-so-obvious) programming subtleties are avoided in favor of clear step-by-step algorithms.

The last section of this chapter deals with a little-used approach to the numerical solution of Laplace's equation, namely, "probabilistic potential theory." This approach is based on a formal analogy between the Monte Carlo statistics of a group of Brownian particles and the discrete form of Laplace's equation.

The resulting algorithm bears almost no resemblance to conventional Laplace's equation algorithms. In many cases, however, the probabilistic potential theory algorithm is the more efficient of the two.

Chapter 10 shows that there are other numerical procedures (besides solving Laplace's equation) that can be used for finding transmission line parameters. Each of these has its own advantages and disadvantages. The procedures treated in this chapter has the principal disadvantages of being only reasonably accurate (2–4% typical) and not being the most efficient in terms of computer processing time. Their main advantage is that they are easily applicable to almost any problem, therefore could be called the most general. For the latter reason this entire chapter is devoted to them.

9.1   TRANSMISSION LINE PARAMETERS
FROM LAPLACE'S EQUATION

It has been shown that for a TEM or quasi-TEM transmission line the capacitance per unit length C is frequency independent. The static or dc capacitance is therefore identically C. Also, for a uniform line a two-dimensional cross section of the line properly and completely describes C.

Consider an arbitrary transmission line cross section, as in Figure 69. A center conductor, set at +1 volt, is surrounded by a 0 volt (grounded) outer conductor. If the outer conductor does not completely enclose the center conductor in an example, assume that a closed path at 0 volts exists at some arbitrary distance that is far from the center conductor. Physically, this assumption is equivalent to requiring that no electric field lines go to infinity.

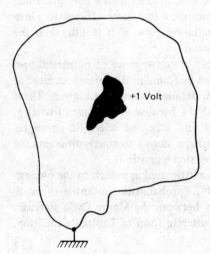

+1 Volt

Figure 69   Arbitrary enclosed transmission line cross section.

If the electric field **E** were known for all the enclosed space, the total energy stored in the electric field (per unit length) U would be given by

$$U = \frac{1}{2} \iint \epsilon \, |\,\mathbf{E}\,|^2 \, dA \tag{9.1}$$

Note that ϵ is possibly $\epsilon(x, y)$, and that the integral is taken over the entire enclosed space.

In terms of circuit parameters, the same stored energy is

$$U = \frac{1}{2} \, CV_0^2 \tag{9.2}$$

where V_0 in this case should be set to 1. Therefore, equating the two equations above, we have

$$C = \iint \epsilon \, |\,\mathbf{E}\,|^2 \, dA \tag{9.3}$$

Alternatively, if the electric field is known over the entire enclosed region, then of course it is known near the surface of the center conductor, where it is normal to the center conductor. At this point all conductors are considered ideal, or nearly so, therefore the normality of static electric field lines at the surface of a conductor has been established. By Gauss' law, the total charge on the center conductor is

$$Q = \int_c \epsilon \mathbf{E} \cdot \mathbf{ds} = \int_c \epsilon E_N \, dl \tag{9.4}$$

where c is the circumference of the center conductor.

Once Q is known, we can write

$$C = \frac{Q}{V_0} = \int_c \epsilon E_N \, dl \tag{9.5}$$

The same argument can be made for the electric field at the surface of the outer conductor. Equation 9.5 will also hold, with the integral being taken over the inward-facing circumference of the outer conductor.

If the voltage $V(x, y)$ is known everywhere in the enclosed region, then **E** is also known everywhere, since

$$\mathbf{E} = -\nabla V(x, y) \tag{9.6}$$

for the static case. The problem of finding C, using (9.3) or (9.5), has been reduced to the problem of finding **E** everywhere in the enclosed region. Using (9.6), this problem has in turn been reduced to that of finding V everywhere in the enclosed region.

Ideal metals are equipotentials, therefore $\mathbf{E} = 0$ on the interior of these metals. The interior of the center conductor is then of no interest in calculating C. In the remaining enclosed region, the dielectric region, V is described by Laplace's equation and the boundary conditions. In a two-dimensional rectangular system, for a uniform dielectric, this is

$$\nabla \cdot \mathbf{D} = \epsilon \nabla \cdot \mathbf{E} = -\epsilon \nabla^2 V(x, y) = 0 \qquad (9.7)$$

or

$$\frac{\partial^2 V}{\partial x^2} + \frac{\partial^2 V}{\partial y^2} = 0 \qquad (9.8)$$

9.2 THE DISCRETE FORM OF LAPLACE'S EQUATION

To solve Laplace's equation numerically, it is necessary to approximate it by a set of discrete difference equations. This means that a set of grid points will be imposed on the transmission line cross section, and the voltage will be considered only at these points. For the sake of simplicity, assume a uniform square grid. That is, let

$$\Delta x = x_{i+1} - x_i = x_i - x_{i-1} = \Delta y = y_{i+1} - y_i = y_i - y_{i-1} \equiv \Delta \qquad (9.9)$$

If the enclosed space is w units wide by b units high, then i and j have maximum values given by

$$I_{max} = 1 + \frac{w}{\Delta} \qquad (9.10a)$$

$$J_{max} = 1 + \frac{b}{\Delta} \qquad (9.10b)$$

Note that we have adopted the FORTRAN convention that subscripts (indices) start at 1.

At the point (i, j) the voltage is denoted $V_{i,j}$. Along the line connecting (i, j) and $(i + 1, j)$,

$$\frac{\partial V}{\partial x} \simeq \frac{V_{i+1,j} - V_{i,j}}{\Delta} \qquad (9.11)$$

Similarly, along the line connecting $(i - 1, j)$ and (i, j),

$$\frac{\partial V}{\partial x} \simeq \frac{V_{i,j} - V_{i-1,j}}{\Delta} \qquad (9.12)$$

The second derivative at (i, j) is then approximated as

$$\frac{\partial^2 V}{\partial x^2} \simeq \frac{(V_{i+1,j} - V_{i,j})/\Delta - (V_{i,j} - V_{i-1,j})/\Delta}{\Delta} = \frac{V_{i+1,j} + V_{i-1,j} - 2V_{i,j}}{\Delta^2} \quad (9.13)$$

Similarly,

$$\frac{\partial^2 V}{\partial y^2} \simeq \frac{V_{i,j+1} + V_{i,j-1} - 2V_{i,j}}{\Delta^2} \quad (9.14)$$

The discrete form of Laplace's equation in the plane is obtained by setting the sum of the two equations above equal to zero:

$$V_{i+1,j} + V_{i-1,j} + V_{i,j+1} + V_{i,j-1} - 4V_{i,j} = 0 \quad (9.15)$$

The importance of (9.15) is that it allows a second-order partial differential equation to be approximated by a set of linear algebraic equations. Although in principle the latter is usually easier to solve than the former, a brute force approach is almost always ill advised. This is because a formal solution requires the inversion of an $(I_{max} \cdot J_{max}) \times (I_{max} \cdot J_{max})$ matrix. Obviously for this approach to be of any practical value, an efficient solution procedure must be developed.

Fortunately in the case of a set of linear equations developed from the discrete form of Laplace's equation, such a procedure exists. This procedure, known as the Gauss-Seidel iteration procedure, has two main advantages: it efficiently converges to the desired solution, and no intermediate calculations must be saved other than the grid voltages themselves.

The Gauss-Seidel iteration procedure is a standard numerical procedure for solving sets of algebraic equations. Consider the following three unknown set of linear equations, for example:

$$2x + y + z = 7$$
$$x - 3y - z = -8 \quad (9.16)$$
$$x + y + 4z = 15$$

These equations must first be rewritten—solving for each of the variables in terms of the other variables, in order:

$$x = \frac{7 - y - z}{2}$$

$$y = \frac{x - z + 8}{3} \quad (9.17)$$

$$z = \frac{15 - x - y}{4}$$

The solution procedure is to repeatedly solve (9.17) in order, starting with an arbitrary set of initial values, and using the latest calculation of each value at all times. For the example above, starting with $(0, 0, 0)$, the calculations would be

$$x = \frac{7 - 0 - 0}{2} = 3.50$$

$$y = \frac{3.50 - 0 + 8}{3} = 3.83$$

$$z = \frac{15 - 3.50 - 3.83}{4} = 1.92 \tag{9.18}$$

$$x = \frac{7 - 3.83 - 1.92}{2} = 0.62$$

etc.

Table 3 shows the results of 6 iterations. As indicated, the solution set has converged to three significant figures.

It should be noted that as a general solution procedure, the Gauss-Seidel iterates do not always converge. If (9.16) were written in a different order, for example,

$$x = 15 - y - 4z$$

$$y = 7 - 2x - z \tag{9.19}$$

$$z = x + 3y + 8$$

the iterates would diverge, regardless of the initial values. Table 4 shows the initial values $(0, 0, 0)$ applied to equations (9.19).

A sufficient convergence criterion for Gauss-Seidel iterates for a set of linear

**Table 3   Example of a Convergent
Gauss-Seidel Iteration**

X	Y	Z	
0	0	0	$X = \dfrac{7 - Y - Z}{2}$
3.50	3.83	1.92	
0.63	2.24	3.03	
0.87	1.95	3.05	$Y = \dfrac{X - Z + 8}{3}$
1.00	1.98	3.01	
1.01	2.00	3.00	$Z = \dfrac{5 - X - Y}{4}$
1.00	2.00	3.00	

Table 4 Example of a Divergent Gauss-Seidel Iteration

X	Y	Z	
0	0	0	
15.0	-23.0	-61.0	$X = 15 - Y - 4Z$
282.	-496.	-1,198.	$Y = 7 - 2X - Z$
5303.	-9401.	-22,892.	$Z = X + 3Y + 8$

equations is that the absolute value of the coefficient of the variable placed on the left-hand side [(9.17) or (9.19)], be as least as large as the sum of the absolute values of the coefficients of all the other variables in each equation. Inspection of (9.15) shows that this condition is satisfied for each equation in the discrete Laplace's equation if (9.15) is solved for $V_{i,j}$:

$$V_{i,j} = \frac{V_{i+1,j} + V_{i-1,j} + V_{i,j+1} + V_{i,j-1}}{4} \qquad (9.20)$$

Thus the efficient solution procedure necessary for practical solutions of the discrete Laplace's equation can always be the Gauss-Seidel iteration procedure. The iteration algorithm, simply stated, is to repeatedly "walk" through the matrix of points, replacing each grid point voltage with the average of the voltages on the four points closest to it. Grid points lying at metal surfaces are of course fixed to the applied voltages, and represent boundary conditions.

9.3 THE TOTAL ENERGY CAPACITANCE CALCULATION

Assume that the correct grid voltages $V_{i,j}$ are known over an entire grid. For the elementary square area taken anywhere in the grid, as in Figure 70, the approximate value of E_x can be found by averaging the values obtained from numerical derivatives taken at j and $j + 1$,

$$E_x \simeq \frac{-1}{2} \left[\frac{V_{i+1,j} - V_{i,j}}{\Delta} + \frac{V_{i+1,j+1} - V_{i,j+1}}{\Delta} \right]$$

$$= \frac{-1}{2\Delta} [V_{i,j+1} - V_{i,j} + V_{i+1,j+1} - V_{i+1,j}] \qquad (9.21)$$

Similarly, for E_y

$$E_y \simeq \frac{-1}{2\Delta} [V_{i,j+1} - V_{i,j} + V_{i+1,j+1} - V_{i+1,j}] \qquad (9.22)$$

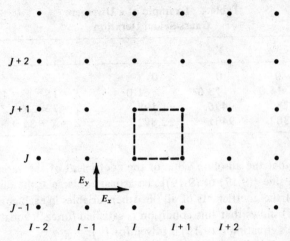

Figure 70 Elementary square grid for stored energy calculations.

The energy stored in the electric field in this area is

$$\Delta U = \tfrac{1}{2} \epsilon \, (E_x^2 + E_y^2) \, \Delta^2 \tag{9.23}$$

And, by direct substitution,

$$\Delta U = \frac{\epsilon}{4} \, [(V_{i,j} - V_{i+1,j+1})^2 + (V_{i+1,j} - V_{i,j+1})^2] \tag{9.24}$$

The total energy stored in the grid is the sum of ΔU over all the elementary squares in the grid,

$$U = \sum_{i=1}^{I_{\max}-1} \sum_{j=1}^{J_{\max}-1} \Delta U \tag{9.25}$$

Using (9.2), with $V_0 = 1$, we write

$$C = 2U \tag{9.26}$$

9.4 A SIMPLE RELAXATION PROGRAM

Example 1. Consider the stripline cross section appearing in Figure 71. The outer conductor is rectangular, 99 units wide and 49 units high. Therefore the enclosed area, taking grid point separations of 1 unit, is

$$1 \leqslant i \leqslant 100$$
$$1 \leqslant j \leqslant 50 \tag{9.27}$$

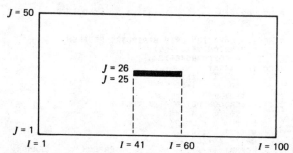

Figure 71 Cross section of transmission line for relaxation grid programming examples.

The center conductor is centered in the box, and is 19 units wide and 1 unit high:

$$41 \leqslant i \leqslant 60$$
$$25 \leqslant j \leqslant 26 \tag{9.28}$$

(*Aside.* The dimensions of C are farads per meter in m.k.s. units. This is identically the dimensions of ϵ, which must appear explicitly as a scale factor in every capacitance calculation. Therefore the cross-sectional dimensions of the transmission line must cancel, and only the ratios are relevant in finding C. Because of this, it is not necessary to specify any units when describing the geometry for a particular calculation.)

The array $V_{i,j}$ consists of $100 \times 50 = 5000$ voltages. This means that each iteration will calculate almost 5000 new values—the "almost" arising because boundary conditions are set and never need to be recalculated. To monitor convergence, some error criterion must be defined. Clearly, it is impractical to inspect almost 5000 numbers manually after each iteration.

Two easily definable error criteria are calculated in the computer program described below. At each calculation using (9.20), the new value of $V_{i,j}$, denoted VNEW, is compared to the old value. The square of the difference between the two is defined to be the local error, ERR. In FORTRAN notation, we write

$$ERR = (VNEW - V(I, J))**2 \tag{9.29}$$

As the iteration proceeds through all i and j, the largest value of ERR is saved and labeled ERRMAX for that iteration. Also, all values of ERR are summed and the sum is labeled ERRSUM for that iteration. The iteration count itself is labeled NRPASS. As $V_{i,j}$ converges to the correct values, we would expect both ERRMAX and ERRSUM to go to zero.

```
FORTRAN IV-PLUS V02-51D           16:43:53     05-OCT-78           PAGE 1
EX1.FTN              /TR:ALL/WR

          C  SIMPLE RELAXATION GRID STRIPLINE PROBLEM
0001               DIMENSION V(100,50)
0002               DATA EPS/8.854E-12/
          C  ZERO THE ARRAY
0003               DO 10  I = 1,100
0004               DO 10  J = 1,50
0005          10   V(I,J) = 0.
          C  SET CENTER CONDUCTOR AT 1 VOLT
0006               DO 20  I = 41,60
0007               DO 20  J = 25,26
0008          20   V(I,J) = 1.0
0009               WRITE (5,1000)
0010        1000   FORMAT (///,'1   NR          ERRSUM        ERRMAX        C',/)
          C  INITIALZE ERROR MONITORS AND START ITERATION LOOP
0011               DO 200  NRPASS = 1,100
0012               ERRMAX = 0.
0013               ERRSUM = 0.
          C  SET UP RELAXATION LOOP
0014               DO 100  I = 2,99
0015               DO 100  J = 2,49
0016               IF (V(I,J) .EQ. 1.)  GO TO 100
0017               VNEW = (V(I+1,J) + V(I-1,J) + V(I,J+1) + V(I,J-1))/4.
0018               ERR = (VNEW - V(I,J))**2
0019               IF (ERR .GT. ERRMAX) ERRMAX = ERR
0020               ERRSUM = ERRSUM + ERR
0021               V(I,J) = VNEW
0022         100   CONTINUE
          C  LIST THE RESULTS OF EVERY TENTH PASS
0023               IF ((NRPASS/10)*10 - NRPASS)  200, 140, 200
0024         140   C1 = 0.
          C  CALCULATE STORED ENERGY CAPACITANCE
0025               DO 150 I = 1,99
0026               DO 150 J = 1,49
0027         150   C1 = C1 + (V(I,J) - V(I+1,J+1))**2
             2         + (V(I+1,J) - V(I,J+1))**2
0028               C1 = C1*EPS/2.
0029               WRITE (5,1100)  NRPASS, ERRSUM, ERRMAX, C1
0030        1100   FORMAT (I6,3X,4G12.3,/)
0031         200   CONTINUE
0032               CALL EXIT
0033               END
```

Figure 72   Example 1: source code listing.

Figure 72 is a program listing of this example. Here the capacitance C1 is calculated only every tenth iteration. Also, a line of output, consisting of NRPASS,
ERRSUM, ERRMAX, and C1 is printed only every tenth iteration.

All $V_{i,j}$ are initialized to zero. The boundary condition on the outer conductor is set simply by never recalculating V at the outer conductor. The
boundary condition at the center conductor is set by initializing to 1 all grid
points within or on the edge of the center conductor. Since the iteration process
is generating new grid voltages by averaging numbers whose maximum is unity
and whose minimum is zero, none of the new numbers can ever be equal to 1.
(It can be seen by following through an iteration that the inspection is done
before the iteration, so that there is never a case of four grid voltages equal to 1
being averaged.) Therefore at each value of i and j, the voltage is inspected to see
if it is 1. If it is, the boundary condition is maintained simply by not calculating
a new grid voltage for that point.

NR	ERRSUM	ERRMAX	C
10	0.997E-01	0.802E-03	0.826E-10
20	0.342E-01	0.175E-03	0.639E-10
30	0.184E-01	0.713E-04	0.556E-10
40	0.119E-01	0.373E-04	0.505E-10
50	0.853E-02	0.222E-04	0.471E-10
60	0.648E-02	0.144E-04	0.446E-10
70	0.513E-02	0.994E-05	0.426E-10
80	0.418E-02	0.722E-05	0.410E-10
90	0.347E-02	0.541E-05	0.397E-10
100	0.293E-02	0.418E-05	0.386E-10

Figure 73   Example 1: output listing.

The entire program—excluding comments—is only 33 FORTRAN source statements long. Of these 33 statements, only one (line 17) is involved in actually calculating grid voltages. The rest of the program is concerned with boundary conditions, error monitoring, and finding $C1$. Note that a calculation of $C1$ is a more time-consuming operation than an entire iteration. If this were a "production" program being used to generate tables rather than an example, $C1$ would be calculated only after some convergence criterion had been satisfied.

Inspection of the program output (Figure 73), indicates that the simple relaxation procedure leaves something to be desired. After 100 iterations both ERRMAX and ERRSUM are falling, but their rates of fall are leveling off. Also, C is not clearly converging to any final value. Either more iterations are needed, or a means of quickening convergence must be found.

9.5 OVERRELAXATION AND PREDICTOR EQUATIONS

The rate of convergence of an iterative scheme can be decelerated and often accelerated by feeding back the old values to some extent when calculating new values. In terms of (9.20), let

$$(V_{i,j})_{\text{new}} = R \left[\frac{V_{i+1,j} + V_{i-1,j} + V_{i,j+1} + V_{i,j-1}}{4} \right] + (1 - R)(V_{i,j})_{\text{old}} \quad (9.30)$$

For $R = 1$, this equation is identically (9.20). For $R = 0$ no change in values takes place. For $0 < R < 1$, the iteration is said to be underrelaxed. Depending on the choice of R, the amount of change that $V_{i,j}$ can undergo is limited. Underrelaxation is often used to stabilize marginally stable numerical procedures.

If $R > 1$, the procedure is overrelaxed. For $1 < R < 2$ the numerical form of Laplace's equation is stable and will converge more rapidly than will the simple case. Although no optimum value of R has been determined for the general case, $R = 1.5$ seems to work well in most situations.

Example 2. To add overrelaxation to the previous example, it is necessary to define the damping parameter, DAMP = 1.5, at the top of the program, and also to replace line 17 of Figure 72 with (9.30):

VNEW = (V(I+1,J) + V(I-1,J) + V(I,J+1) + V(I,J-1))/4.*DAMP 2

$$+ (1.-DAMP)*V(I,J) \quad (9.31)$$

Figure 74 gives the listing and output generated by this modified program. The usefulness of overrelaxation is seen by comparing this output with the previous case. The accuracy of the final value of C is still in doubt, however, after 100 iterations. Rather than simply increasing the number of iterations, another convergence accelerating scheme may be devised based on the observation that although most of the computer time is spent calculating values for the various $V_{i,j}$, the actual values of these voltages are not the desired result. The desired result is the correct value of C. Perhaps computer time could be profitably used performing calculations directly on the intermediate values of C.

Inspection of the output listing in Figure 74 reveals that C is converging monotonically to some final value. If the intermediate values of C as a function of NRPASS could be curve-fitted to some asymptotic function, it might be possible to predict the final value of C without having iterated enough times to calculate C accurately.

Example 3. The choice of such a curve-fitting function is not a well-defined problem, and there is no assurance that an arbitrarily chosen function will be optimum in any sense. As an example of a function, consider a simple two-point function of the form

$$C(N) = \frac{A}{1 + (B/N)^M} \quad (9.32)$$

where $N = $ NRPASS, the iteration counter
$M = $ a positive number
$A, B = $ coefficients determined from two consecutive calculations of $C(N)$

As N gets large, the equation approaches A. In the limit as N goes to infinity, A is the exact value of C. For $C_1 = C(N_1)$ and $C_2 = C(N_2)$, by direct substitution into (9.32) and elimination of B,

$$A = \frac{N_2^M - N_1^M}{(N_2/C_2)^M - (N_1/C_1)^M} \quad (9.33)$$

Figure 75 is a listing of the stripline calculation computer program, including both overrelaxation and the "predictor" equation (9.33). In this example $M = 3$.

```
        C  SIMPLE RELAXATION GRID STRIPLINE PROBLEM
0001           DIMENSION V(100,50)
0002           DATA EPS/8.854E-12/
        C  ZERO THE ARRAY
0003           DO 10  I = 1,100
0004           DO 10  J = 1,50
0005      10   V(I,J) = 0.
        C  SET CENTER CONDUCTOR AT 1 VOLT
0006           DO 20  I = 41,60
0007           DO 20  J = 25,26
0008      20   V(I,J) = 1.0
        C  SET THE OVERRELAXATION PARAMETER
0009           DAMP = 1.5
0010           WRITE (5,1000)
0011    1000   FORMAT (//,'1   NR          ERRSUM       ERRMAX       C',/)
        C  INITIALZE ERROR MONITORS AND START ITERATION LOOP
0012           DO 200  NRPASS = 1,100
0013           ERRMAX = 0.
0014           ERRSUM = 0.
        C  SET UP RELAXATION LOOP
0015           DO 100  I = 2,99
0016           DO 100  J = 2,49
0017           IF (V(I,J) .EQ. 1.)  GO TO 100
0018           VNEW = (V(I+1,J) + V(I-1,J) + V(I,J+1) + V(I,J-1))/4.
          2       *DAMP + (1. - DAMP)*V(I,J)
0019           ERR = (VNEW - V(I,J))**2
0020           IF (ERR .GT. ERRMAX) ERRMAX = ERR
0021           ERRSUM = ERRSUM + ERR
0022           V(I,J) = VNEW
0023     100   CONTINUE
        C  LIST THE RESULTS OF EVERY TENTH PASS
0024           IF ((NRPASS/10)*10 - NRPASS) 200, 140, 200
0025     140   C1 = 0.
        C  CALCULATE STORED ENERGY CAPACITANCE
0026           DO 150 I = 1,99
0027           DO 150 J = 1,49
0028     150   C1 = C1 + (V(I,J) - V(I+1,J+1))**2
          2         + (V(I+1,J) - V(I,J+1))**2
0029           C1 = C1*EPS/2.
0030           WRITE (5,1100)  NRPASS, ERRSUM, ERRMAX, C1
0031    1100   FORMAT (I6,3X,4G12.3,/)
0032     200   CONTINUE
0033           CALL EXIT
0034           END
```

(a)

NR	ERRSUM	ERRMAX	C
10	0.190	0.110E-02	0.567E-10
20	0.623E-01	0.189E-03	0.450E-10
30	0.328E-01	0.656E-04	0.400E-10
40	0.202E-01	0.311E-04	0.371E-10
50	0.131E-01	0.167E-04	0.352E-10
60	0.886E-02	0.960E-05	0.340E-10
70	0.613E-02	0.573E-05	0.332E-10
80	0.433E-02	0.371E-05	0.326E-10
90	0.312E-02	0.262E-05	0.322E-10
100	0.229E-02	0.194E-05	0.319E-10

(b)

Figure 74   Example 2: (*a*) source code and (*b*) output listing.

149

```
      C  SIMPLE RELAXATION GRID STRIPLINE PROBLEM
      C  OVERRELAXATION AND PREDICTION ADDED
0001         DIMENSION V(100,50)
0002         DATA U,EPS,XN1,Y2/12.57E-7,8.854E-12,0.,1./
      C  ZERO THE ARRAY
0003         DO 10  I = 1,100
0004         DO 10  J = 1,50
0005     10  V(I,J) = 0.
      C  SET CENTER CONDUCTOR AT 1 VOLT
0006         DO 20  I = 41,60
0007         DO 20  J = 25,26
0008     20  V(I,J) = 1.0
      C  SET THE OVERRELAXATION PARAMETER
0009         DAMP = 1.5
0010         WRITE (5,1000)
0011   1000  FORMAT (/,'1   NR     ERRSUM      ERRMAX       C: CALC - - PRED
      2  ',/)
      C  INITIALZE ERROR MONITORS AND START ITERATION LOOP
0012         DO 200  NRPASS = 1,100
0013         ERRMAX = 0.
0014         ERRSUM = 0.
      C  SET UP RELAXATION LOOP
0015         DO 100  I = 2,99
0016         DO 100  J = 2,49
0017         IF (V(I,J) .EQ. 1.)  GO TO 100
0018         VNEW = (V(I+1,J) + V(I-1,J) + V(I,J+1) + V(I,J-1))/4.
      2         *DAMP + (1. - DAMP)*V(I,J)
0019         ERR = (VNEW - V(I,J))**2
0020         IF (ERR .GT. ERRMAX) ERRMAX = ERR
0021         ERRSUM = ERRSUM + ERR
0022         V(I,J) = VNEW
0023   100   CONTINUE
      C  LIST THE RESULTS OF EVERY TENTH PASS
0024         IF ((NRPASS/10)*10 - NRPASS)  200, 140, 200
0025   140   C1 = 0.
      C  CALCULATE STORED ENERGY CAPACITANCE
0026         DO 150 I = 1,99
0027         DO 150 J = 1,49
0028   150   C1 = C1 + (V(I,J) - V(I+1,J+1))**2
      2       + (V(I+1,J) - V(I,J+1))**2
0029         C1 = C1*EPS/2.
      C  PREDICT CAPACITANCE
0030         XN1 = XN2
0031         XN2 = FLOAT(NRPASS)**3
0032         Y1 = Y2
0033         Y2 = C1
0034         C = CFIN
0035         CFIN = (XN2-XN1)/(XN2/Y2 - XN1/Y1)
      C  CAPACITANCE CANNOT BE PREDICTED BEFORE TWO CALCULATIONS
0036         IF (NRPASS .LT. 20)  GO TO 200
0037         WRITE (5,1100) NRPASS, ERRSUM, ERRMAX, C1, CFIN
0038   1100  FORMAT (I6,3X,5G12.3,/)
0039         IF (ABS(C-CFIN) .LT. .01*ABS(C)) GO TO 250
0040   200   CONTINUE
0041   250   Z0 = SQRT(U*EPS)/C
0042         WRITE (5,1200) Z0
```

(a)

```
0043   1200  FORMAT (///,' THE LINE IMPEDANCE IS',F5.0,' OHMS ')
0044         CALL EXIT
0045         END
```

(b)

Figure 75 Example 3: listing and output.

150

```
 NR      ERRSUM      ERRMAX      C: CALC - - PRED

 20     0.623E-01   0.189E-03   0.450E-10   0.437E-10
 30     0.328E-01   0.656E-04   0.400E-10   0.382E-10
 40     0.202E-01   0.311E-04   0.371E-10   0.352E-10
 50     0.131E-01   0.167E-04   0.352E-10   0.335E-10
 60     0.886E-02   0.960E-05   0.340E-10   0.325E-10
 70     0.613E-02   0.573E-05   0.332E-10   0.319E-10
 80     0.433E-02   0.371E-05   0.326E-10   0.315E-10
 90     0.312E-02   0.262E-05   0.322E-10   0.313E-10
```

THE LINE IMPEDANCE IS 106. OHMS
>

(c)

Figure 75 *(Continued)*

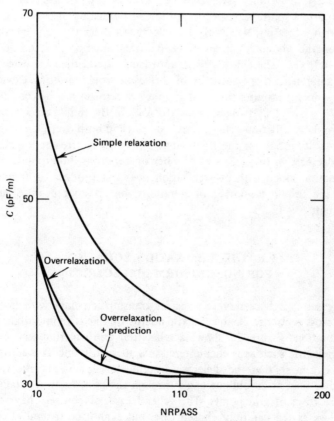

Figure 76 *C* versus NRPASS for Examples 1, 2, and 3.

To make the program more generally useful, the following features were also added:

1. The predicted values of C are monitored, and the iterations are terminated when two successive predictions agree to within 1%. Note that since C is calculated only every tenth iteration, the predictor equation is applied only every tenth iteration.

2. Once C is known, Z_0 is calculated—assuming that $\epsilon_r = 1$.

As the output listing indicates, the predicted value of C has converged sufficiently to satisfy the 1% criterion after only 90 iterations.

Examples 2 and 3 have also been solved analytically (see the references at the end of Chapter 6). According to the analytic results, the correct answer to this problem is $Z_0 = 109$ ohms. The calculated value of 106 ohms is therefore within 3% of the correct solution. Inspection of the output listing shows that C is falling in value with increasing NRPASS. If a tolerance tighter than 1% agreement was required on the predicted values of C, the final value of C would decrease by approximately 2%. The characteristic impedance would then increase by about the same amount, and the accuracy of the result would improve. Obviously the tradeoff between computer time and accuracy of results must be considered.

Figure 76 shows C as a function of NRPASS for the simple relaxation calculation, the same calculation using overrelaxation, and both overrelaxation and the predictor equation. From the figure it is clear that overrelaxation results in a dramatic decrease in the necessary number of iterations. Incorporating the predictor equation along with overrelaxation does not produce as dramatic a decrease in the required number of iterations, but it does cause a noticeable improvement.

9.6 THE RELAXATION EQUATION FOR NONUNIFORM DIELECTRICS

In many cases it is necessary to consider transmission lines with nonuniform dielectric cross sections. The quasi-static microstrip line is an important example of this situation. Figure 77 shows a relaxation grid superimposed on a nonuniform dielectric such as is encountered in a microstrip line. It is convenient to arrange the grid so that the dielectric interface is along a specific value of J (or I), rather than at an oblique angle through the grid, or even parallel to but in between rows of grid points. The general case, of course, obeys the same physical laws as the carefully chosen case, but cannot be treated as concisely. Numerical treatment of the general case is discussed in the references at the end of this chapter.

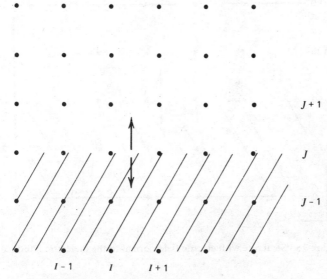

Figure 77 Cross section of a relaxation grid superimposed on a nonuniform dielectric.

For the situation shown in Figure 77, above the dielectric interface,

$$D_y \simeq \frac{-\epsilon_0 \epsilon_{r1}(V_{i,j+1} - V_{i,j})}{\Delta} \qquad (9.34)$$

and below the interface,

$$D_y \simeq \frac{-\epsilon_0 \epsilon_{r2}(V_{i,j} - V_{i,j-1})}{\Delta} \qquad (9.35)$$

At the dielectric interface, therefore,

$$\frac{\partial D_y}{\partial y} \simeq \frac{-\epsilon_0}{\Delta} [\epsilon_{r1}(V_{i,j+1} - V_{i,j}) - \epsilon_{r2}(V_{i,j} - V_{i,j-1})] \qquad (9.36)$$

To write an iteration equation for $V_{i,j}$ along the interface, a relation for $\partial D_x/\partial x$ at the interface is also needed. It is not obvious what value to use for ϵ_r at $(i - 1, j)$ or $(i + 1, j)$—although whatever value is used must be the same at both points. For the moment, refer to the required value as ϵ_{r3}. Then $\partial D_x/\partial x$ is

$$\frac{\partial D_x}{\partial x} \simeq \frac{-\epsilon_0 \epsilon_{r3}}{\Delta} [V_{i+1,j} + V_{i-1,j} - 2V_{i,j}] \qquad (9.37)$$

and along the interface the required iteration equation is

$$V_{i,j} = \frac{\epsilon_{r1} V_{i,j+1} + \epsilon_{r2} V_{i,j-1} + \epsilon_{r3}(V_{i-1,j} + V_{i+1,j})}{\epsilon_{r1} + \epsilon_{r2} + 2\epsilon_{r3}} \qquad (9.38)$$

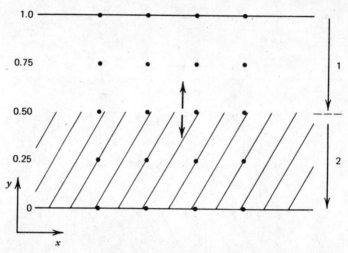

Figure 78   Structure with electric field normal to the dielectric interface.

Note that (9.38) is necessary only along the interface, and that the basic iteration equation, (9.20), holds both above and below the interface.

To determine the correct value of ϵ_{r3}, consider an example from electrostatics that can be solved exactly—the cross section of an infinitely wide pair of plates. The plates are 1 unit apart, with 0 volts on the lower plate and 1 volt on the upper plate (Figure 78). Let the lower half of the space between the plates be a dielectric with relative constant ϵ_{r2}, and the upper half of the space be a dielectric with relative constant ϵ_{r1}.

By inspection, $E_x = D_x = 0$, and

$$\nabla \cdot \mathbf{D} = \frac{\partial D_y}{\partial y} = 0 \tag{9.39}$$

The value of D_y must then be constant throughout the dielectrics. For E_1 and E_2 the electric fields in the upper and lower half-regions, respectively, since D_y is constant,

$$\epsilon_{r1} E_1 = \epsilon_{r2} E_2 \tag{9.40}$$

The applied voltage between the plates is the line integral of the fields, and since E_1 and E_2 are themselves constants,

$$V = 1 = -\int_0^{1/2} E_2 \, dy - \int_{1/2}^1 E_1 \, dy = -\frac{E_1 + E_2}{2} \tag{9.41}$$

or

$$E_1 + E_2 = -2 \qquad (9.42)$$

Solving (9.40) and (9.42) simultaneously, we have

$$E_1 = \frac{-2\epsilon_{r2}}{\epsilon_{r1} + \epsilon_{r2}} \qquad (9.43a)$$

$$E_2 = \frac{-\epsilon_{r1}}{\epsilon_{r1} + \epsilon_{r2}} \qquad (9.43b)$$

The voltages at the grid points now can be evaluated:

$$V(0.25) = -0.25E_2 = -\frac{0.5\epsilon_1}{\epsilon_1 + \epsilon_2}$$

$$V(0.5) = -0.5E_2 = -\frac{\epsilon_1}{\epsilon_1 + \epsilon_2} \qquad (9.44)$$

$$V(0.75) = -0.5E_2 - 0.25E_1 = -\frac{\epsilon_1 + 0.5\epsilon_2}{\epsilon_1 + \epsilon_2}$$

Substituting these values into (9.38) yields

$$V(0.5) = \epsilon_1 V(0.75) + \epsilon_2 V(0.25) + 2\epsilon_3 V(0.5) \qquad (9.45)$$

or

$$\epsilon_1 + \epsilon_2 + 2\epsilon_3 = \epsilon_1 + \epsilon_2 + 2\epsilon_3 \qquad (9.46)$$

It would appear that, for y-directed fields, therefore for the y components of any electric fields, the choice of ϵ_3 is irrelevant.

Now consider the case depicted in Figure 79. In this case two parallel plates, infinite in extent, are both parallel to the y axis. As before, the plates are 1 unit apart and have a potential difference of 1 volt. The electric field is therefore in the x direction. The dielectric interface in this case, however, is parallel to the fields. Since the (x-directed) field lines do not cross the dielectric interface, the voltage at any point x is directly proportional to the distance of the point from the left-hand plate. Using (9.38), then,

$$V(0.5) = 0.5 = \frac{0.5\epsilon_1 + 0.5\epsilon_2 + (0.75 + 0.25)\epsilon_3}{\epsilon_1 + \epsilon_2 + 2\epsilon_3} = 0.5 \left[\frac{\epsilon_1 + \epsilon_2 + 2\epsilon_3}{\epsilon_1 + \epsilon_2 + 2\epsilon_3} \right] \qquad (9.47)$$

Again, the value of ϵ_3 is irrelevant.

Since the laws of electrostatics offer no guidance in a choice of ϵ_3, perhaps the numerical procedures can suggest some possibilities for optimizing the convergence time and efficiency. A choice of ϵ_3 that would cause the $V_{i+1,j}$ and $V_{i-1,j}$ terms to contribute an equal amount as the $V_{i,j-1}$ and the $V_{i,j+1}$ terms, therefore assuring a reasonable convergence at the interface, would be, using

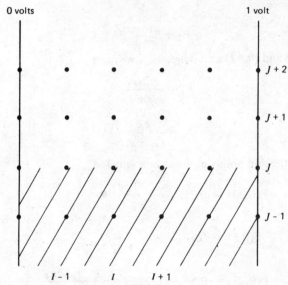

Figure 79 Structure with electric field parallel to the dielectric interface.

equation (9.38)

$$\epsilon_1 V_{i,j+1} + \epsilon_2 V_{i,j-1} = \epsilon_3 (V_{i+1,j} + V_{i-1,j}) \tag{9.48a}$$

or

$$\epsilon_3 = \frac{\epsilon_1 V_{i,j+1} + \epsilon_2 V_{i,j-1}}{V_{i+1,j} + V_{i-1,j}} \tag{9.48b}$$

Unfortunately, this choice requires an updating of ϵ_3 using the most recent voltage values at each calculation along the dielectric interface. This is therefore very inefficient. Making the observation that the four voltages involved in (9.48) are probably not very different from one another, let

$$\epsilon_3 = \frac{\epsilon_1 + \epsilon_2}{2} \tag{9.49}$$

be an approximation to (9.48b). This choice has the advantage that it is calculated once, outside of all loops in the program.

9.7 THE BOXED MICROSTRIP LINE

Example 4. Figure 80 shows a typical boxed (enclosed) microstrip line. A center conductor 10 units wide and 1 unit thick is situated on top of a dielectric

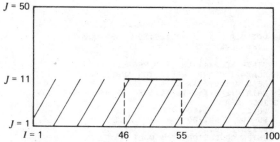

Figure 80   Cross section of boxed microstrip example.

with relative dielectric constant = 10, 10 units thick. The box itself is 99 units wide and 49 units high.

Chapter 5 explained why the characteristic impedance of a quasi-static microstrip line cannot be found from a single capacitance calculation. To find L, it is necessary to calculate the capacitance of the structure with $\epsilon_r = 1$. The program listed in Figure 81 does both calculations.

From a computational point of view, the microstrip problem requires two complete computer runs. This is done automatically in the program as described. The overrelaxation and predictor equations are identical to those described earlier. The output of the program (Figure 81b) reflects the fact that two complete runs are being made. It is obvious from the capacitance values shown that the capacitance with the dielectric material present is found first.

The characteristic impedance for this line, predicted using (5.3) is 45.4 ohms. Therefore using the 1% repeatability criterion for termination of iterations, the program has again produced a characteristic impedance approximately 3% away from the correct answer. The comments made previously about decreasing the allowed variation in the predicted value of C to increase accuracy pertain equally here.

Is it necessary to use an iterative numerical scheme to find a line impedance that can be found from a closed form formula? The answer to this question is that the choice of example for the numerical procedure was made specifically to be one that could be verified easily. The computer program just as easily could have found the impedance of an off-centered microstrip line in a box, a multi-layered dielectric (and/or off-centered) line in a box, or even a microstrip line with the upper conductor shaped like a triangular wedge or a pentagon in cross section.

The point is that simple formulas exist only for situations with high degrees of symmetry and very few complications. The numerical procedure is totally indifferent to the symmetries, the number of dielectrics, and other such variables.

```
        C  MICROSTRIP RELAXATION GRID PROBLEM
        C  OVERRELAXATION AND PREDICTION ADDED
0001           REAL V(100,50), L
0002           DATA U,EPS,XN1,Y2/12.57E-7,8.854E-12,0.,,1./
        C  ZERO THE ARRAY
0003           DO 10   I = 1,100
0004           DO 10   J = 1,50
0005     10    V(I,J) = 0.
        C  SET CENTER CONDUCTOR AT 1 VOLT
0006           DO 20   I = 46,55
0007           DO 20   J = 11,12
0008     20    V(I,J) = 1.0
        C  SET THE OVERRELAXATION PARAMETER
0009           DAMP = 1.5
        C  SET RELATIVE DIELECTRIC AND INTERFACE DIELECTRIC
0010           ER = 10.
0011           E3 = (1.+ER)/2.
0012           WRITE (5,1000)
0013   1000    FORMAT (/,'1     NR     ERRSUM      ERRMAX       C: CALC - - PRED
       2    ',/)
        C  INITIALZE ERROR MONITORS AND START ITERATION LOOP
0014           DO 300   IER = 1,2
0015           IF (IER .EQ. 2)  ER = 1.
0016           DO 200   NRPASS = 1,250
0017           ERRMAX = 0.
0018           ERRSUM = 0.
        C  SET UP RELAXATION LOOP
0019           DO 100   I = 2,99
0020           DO 100   J = 2,49
0021           IF (V(I,J) .EQ. 1.)  GO TO 100
        C  TOP OF DIELECTRIC IS AT J = 11
0022           IF (J - 11)  60, 40, 60
0023     40    VNEW = (V(I,J+1)+ER*V(I,J-1) + E3*(V(I+1,J)+V(I-1,J)))
       2    /(1.+ER+2.*E3)*DAMP + (1.-DAMP)*V(I,J)
0024           GO TO 70
0025     60    VNEW = (V(I+1,J)+V(I-1,J)+V(I,J+1)+V(I,J-1))/4.*DAMP
       2    + (1.-DAMP)*V(I,J)
0026     70    ERR = (VNEW - V(I,J))**2
0027           IF (ERR .GT. ERRMAX) ERRMAX = ERR
0028           ERRSUM = ERRSUM + ERR
0029           V(I,J) = VNEW
0030    100    CONTINUE
        C  LIST THE RESULTS OF EVERY TENTH PASS
0031           IF ((NRPASS/10)*10 - NRPASS)  200, 140, 200
0032    140    C1 = 0.
        C  CALCULATE STORED ENERGY CAPACITANCE
0033           DO 150 I = 1,99
0034           DO 150 J = 1,49
0035           DELC1 = (V(I,J)-V(I+1,J+1))**2 + (V(I+1,J) -
       2    V(I,J+1))**2
0036           IF (J .LT. 11)  DELC1 = DELC1*ER
0037    150    C1 = C1 + DELC1
0038           C1 = C1*EPS/2.
        C  PREDICT CAPACITANCE
0039           XN1 = XN2
0040           XN2 = FLOAT(NRPASS)**3
```

(a)

Figure 81 Example 4 (boxed microstrip) listings.

158

```
FORTRAN-IV-PLUS V02-51D        21:25:50    27-NOV-78         PAGE 2
EX4.FTN              /TR:ALL/WR

0041                  Y1 = Y2
0042                  Y2 = C1
0043                  C = CFIN
0044                  CFIN = (XN2-XN1)/(XN2/Y2 - XN1/Y1)
      C  CAPACITANCE CANNOT BE PREDICTED BEFORE TWO CALCULATIONS
0045                  IF (NRPASS .LT. 20)  GO TO 200
0046                  WRITE (5,1100)  NRPASS, ERRSUM, ERRMAX, C1, CFIN
0047       1100       FORMAT (I6,3X,5G12.3,/)
0048                  IF (ABS(C-CFIN) .LT. .01*ABS(C)) GO TO 250
0049        200       CONTINUE
0050        250       WRITE (5,1200)  NRPASS, CFIN
0051       1200       FORMAT (//,' AFTER',I4,' ITERATIONS, CFIN =',G12.3,/)
0052                  IF (IER -1) 300,300,270
0053        270       L = EPS*U/CFIN
0054                  Z0 = SQRT(L/CSAVE)
0055                  WRITE (5,1300) Z0
0056       1300       FORMAT (' Z0 = ',F8.1,' OHMS')
0057                  CALL EXIT
0058        300       CSAVE = C
0059                  END
```

(b)

```
   NR       ERRSUM        ERRMAX       C: CALC - - PRED

   20      0.204E-01     0.663E-04    0.181E-09   0.178E-09
   30      0.862E-02     0.209E-04    0.177E-09   0.175E-09
   40      0.479E-02     0.103E-04    0.175E-09   0.174E-09


AFTER  40 ITERATIONS, CFIN =    0.174E-09

   20      0.611E-02     0.126E-04    0.302E-10   0.301E-10
   30      0.404E-02     0.574E-05    0.299E-10   0.298E-10
   40      0.293E-02     0.343E-05    0.297E-10   0.296E-10


AFTER  40 ITERATIONS, CFIN =    0.296E-10

Z0 =     46.4 OHMS
>
```

(c)

Figure 81 *(Continued)*

9.8 COUPLED LINES

When two identical lines are coupled, there are two separate parameters to find: Z_{0e} and Z_{0o}. These parameters must be found independently by exciting the system according to the definition of even and odd mode impedances.

As an example, consider the pair of coupled microstrip lines, enclosed in a box, shown in Figure 82. Each line is 10 units wide. The lines are separated by two units. The dielectric is 10 units high and has a relative dielectric constant of 10. The entire system is in a box 199 units wide and 49 units high. The lines themselves are very thin, ideally having zero thickness: in practice this approximation is usually quite reasonable. Using conventional microstrip notation, the normalized width of the lines is $w/h = 1.0$ and the normalized separation of the lines is $s/h = 0.2$.

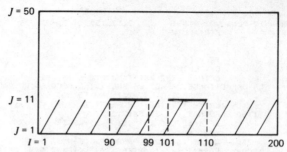

Figure 82 Cross section of coupled microstrip example.

Consider first the odd, or asymmetric, mode. One line has +1 volt applied to it while the other has −1 volt applied to it; Z_{oo} is defined as the impedance measured (or inferred from the capacitance measured) from either line to ground.

Referring to Figure 82, the line $I = 100$ is an equipotential plane, at 0 volts. Therefore Z_{oo} can be calculated by "erasing" everything to the right of $I = 100$ in the figure, and considering the $I = 100$ line as the right-hand enclosing wall of the modified system. In terms of a computer program, the single boxed microstrip line program described in Section 9.7 will solve this problem with the simple modifications of moving the upper conductor toward the right edge and making the upper conductor ideally thin. Specifically, referring to Figure 83, replace computer program statements 11 and 12 by

$$11 \quad DO\ 20 \quad I = 90, 99$$

$$12 \quad J = 11$$

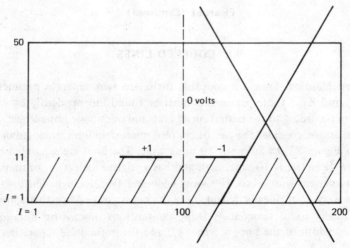

Figure 83 Modification of coupled microstrip for Z_{oo} calculation.

```
       NR.      ERRSUM        ERRMAX        C: CALC - - PRED

       20     0.921E-02     0.641E-04     0.264E-09   0.262E-09
       30     0.326E-02     0.158E-04     0.262E-09   0.261E-09


AFTER  30 ITERATIONS, CFIN =    0.261E-09

       20     0.274E-02     0.133E-04     0.469E-10   0.469E-10
       30     0.159E-02     0.573E-05     0.468E-10   0.468E-10


AFTER  30 ITERATIONS, CFIN =    0.468E-10

ZO =      30.1 OHMS
>
```

Figure 84   Output listing of Z_{0o} calculations.

Since each line must contribute equally to the total stored energy, the value of C resulting from summing up the stored energy in the "half-box" as just described will be the correct capacitance for the line. Therefore no further program modifications are needed. Figure 84 presents the output listing of the modified program.

To calculate Z_{0e} it is necessary to place +1 volt on both lines, then measure the capacitance from either line to ground. In this case the symmetry of the situation makes the line $I = 100$ a "mirror" axis of symmetry—that is, $V_{99,j} = V_{101,j}$, $V_{98,j} = V_{102,j}, \ldots$, for all j. This means that if $V_{i,j}$ is known for all I equal to or less than 100, it is known for all I. Also, as in the asymmetric mode case, the energy stored in half of the box (left or right) will lead to the correct capacitance value.

To modify the program used to find Z_{0o}, it is necessary to create an $I = 101$ column, and, while not iterating through I greater than 100, after each iteration set $V_{101,j} = V_{99,j}$ for all j. Specifically modify the program that calculated Z_{0o} as follows:

1. Change the dimension statement to DIMENSION $V(101,50)$.
2. Change DO 100 I = 2,99 to DO 100 I = 2,100.
3. Immediately after DO 100 J = 2,49, insert $V(101,J) = V(99,J)$.

The output listing of this program as modified appears in Figure 85. Compared with published data, (see references at the end of Chapter 5), both the even and odd mode impedances are correct to better than 3%.

It should be mentioned that the procedures described will work correctly only for identical sections of line. In the case of unequal line widths, the odd mode calculation falls because the equipotential surface between the two lines is not the axis of symmetry. In fact, it is probably not even a straight line. Furthermore, since both lines contribute unequal amounts to the total stored energy of the system, and it is not known a priori how much of the stored energy to ascribe to each line, the individual capacitances cannot be calculated without some additional information.

```
   NR        ERRSUM        ERRMAX        C: CALC - - PRED

   20       0.190E-01     0.754E-04     0.143E-09    0.140E-09
   30       0.813E-02     0.303E-04     0.139E-09    0.138E-09
   40       0.463E-02     0.155E-04     0.138E-09    0.137E-09


AFTER  40 ITERATIONS, CFIN =    0.137E-09

   20       0.432E-02     0.118E-04     0.215E-10    0.214E-10
   30       0.299E-02     0.593E-05     0.212E-10    0.212E-10
   40       0.222E-02     0.395E-05     0.211E-10    0.209E-10


AFTER  40 ITERATIONS, CFIN =    0.209E-10

ZO =     62.1 OHMS
>
```

Figure 85 Output listing of Z_{0e} calculations.

The asymmetric line case may be handled by way of the total energy calculation by setting up the three measurements described in Section 8.5. There is no a priori reason for this procedure to fail to work, but the following problems should be noted:

1. The total calculation time (therefore cost) is large because there are three capacitance calculations required, as opposed to two calculations in the symmetric case. Also, this situation is worsened because symmetry conditions cannot be used to cut down on the required number of grid points, inhomogeneous dielectric problems multiply the required number of calculations by 2, and as reason 2 shows, extreme accuracy is required.

2. In most practical cases the coupling capacitance C_m is about 2 orders of magnitude smaller than either of the individual line capacitances. Since C_m is found using (8.67)—that is, by a subtraction of two nearly equal large numbers to find the small difference between them, relatively small errors in these (large) numbers can easily be larger than the correct result. Proper convergence and sufficient grid point resolution must be carefully assured if this method is to yield any useful results.

9.9 THE TOTAL CHARGE CAPACITANCE

Section 9.1 showed that the capacitance of a line to ground can be found either from the total electrostatic energy stored in the system or from the total charge on the center conductor(s). Using the value of the total charge on each line, the capacitance of each line to ground is found, then the even and odd mode impedances are calculated—noting of course that the proper excitation must be applied for each calculation. A pair of coupled lines that are not identical therefore will have two even and two odd mode impedances, one even and one odd mode impedance for each line. In terms of the capacitive π model, all three capacitors are unequal. In terms of a possible filter design, the source and load impedances for an optimum match are unequal.

Table 5 Voltage Values Around Center Conductor Needed for Total Charge Calculation

I	J = 10	J = 11	J = 12	J = 13
45	0.660	0.730	0.765	0.716
46	0.788	1.000	1.000	0.832
47	0.838	1.000	1.000	0.878
48	0.860	1.000	1.000	0.899
49	0.870	1.000	1.000	0.909
50	0.875	1.000	1.000	0.913
51	0.875	1.000	1.000	0.913
52	0.870	1.000	1.000	0.909
53	0.860	1.000	1.000	0.899
54	0.838	1.000	1.000	0.878
55	0.788	1.000	1.000	0.832
56	0.660	0.730	0.765	0.716

For illustrative purposes, the example below is a repetition of the enclosed single microstrip line problem of Section 9.7. Extending the technique to more than one line is a straightforward procedure, and introducing the technique in a multiline example is an unnecessary complication.

To calculate the electric displacement vector $\mathbf{D}_n$ at the surface of the center conductor, it is necessary to know the voltages at all the grid points surrounding the center conductor. To obtain these, the program for the boxed microstrip (Figure 81) was run for 250 iterations to assume good convergence; then the grid voltages were listed. The voltages in the region of the center conductor appear in Table 5. From these values, the normal electric field strength about the center conductor is calculated. Except at the corners $(i, j = 45, 10, \text{etc})$, since the grid points are 1 unit apart, the electric field is $1 - V_{i,j}$. At the corners,

$$E_x(45, 10) = V_{46, 10} - V_{45, 10}$$

$$E_y(45, 10) = V_{45, 11} - V_{45, 10}$$

etc.

Table 6 shows the results of the field calculations.

To evaluate $\mathbf{D}_n$ each of the E values is multiplied by the appropriate delectric constant. Then Q is obtained from a simple numerical integration—sum up all the D values along each edge except for the first and last values, which contribute one-half of their values to each sum. These sums are then multiplied by the appropriate ϵ_r, and these results themselves are summed. Tabulating this procedure, Table 7 shows the total charges, and finally, $Q = C = 195 \times 10^{-12}$. This capacitance is about 8% larger than the correct value. The principal source of error here is the field calculations at the corners of the center conductor. The

Table 6   Calculated D Vector for Total Charge Calculation

1. Vertical Field Under the Center Conductor ($J = 10$)

I	$E_{i,10}$	I	$E_{i,10}$
45	0.070	46	0.212
47	0.162	48	0.140
49	0.130	50	0.125
51	0.125	52	0.130
53	0.140	54	0.162
55	0.212	56	0.070

2. Vertical Field Above the Center Conductor ($J = 13$)

I	$E_{i,13}$	I	$E_{i,13}$
45	0.049	46	0.168
47	0.122	48	0.101
49	0.091	50	0.087
51	0.087	52	0.091
53	0.101	54	0.122
55	0.168	56	0.049

3. Horizontal Field to the Left (or Right) of the Center Conductor ($I = 45$ or 56)

	J	$E_{i,j}$	J	$E_{i,j}$
10	10	0.128	11	0.270
	12	0.235	13	0.124

electric field lines converge rapidly at the corners, and a very fine grid is necessary to calculate accurately the fields in these regions. The tradeoff is again one of computer time and storage versus desired accuracy. Possible alternatives are to use a nonuniform grid, or to find a scheme that does not require large amounts of storage and unnecessary calculation for detailed study of a particular region of a transmission line cross section.

Table 7   Calculated Total Charge Around Center Condition

$Q_{\text{lower}} = \epsilon_0(10)(0.035 + 0.212 + 0.162 + 0.140 + 0.130 + 0.125 + 0.130$
$\qquad + 0.140 + 0.162 + 0.212 + 0.035) = 142 \times 10^{-12}$

$Q_{\text{upper}} = 10.5 \times 10^{-12}$

$Q_{\text{left}} = Q_{\text{right}} = \epsilon_0(0.064 \times 10 + 0.270 \times 5.5 + 0.235 + 0.062) = 21.4 \times 10^{-12}$

$Q_{\text{total}} = C = 195 \times 10^{-12}$

Assuming that the accuracy obtained eventually is satisfactory, the calculation of the parameters of coupled lines by the technique above requires first the inclusion of all such lines in a relaxation grid computation. The total charge on each line is then found, and the capacitances of the lines are known at that point. Note that in the case of three or more coupled lines, all combinations of excitations must be found. For example, a three-line system has excitations of the form +++, ++-, and +-+. Each excitation yields a corresponding capacitance and impedance. If the lines are unequal, there are three excitation modes leading to three capacitances per line.

9.10 PROBABILISTIC POTENTIAL THEORY

Monte Carlo methods for the solution of partial differential equations have appeared in the literature for more than 30 years. Still, the method is virtually unused by engineers—probably because the method is so far removed from the conventional approaches that it is never discovered by workers interested in transmission line problems. As is shown below, however, the advantages of the method are indisputable in certain cases.

Forget for the moment that there is any interest in transmission line problems, or electrical problems of any sort. Treat the relaxation grid described throughout this chapter as a sort of a "board game," with the following rules of play:

1. A "player" is placed on any of the grid points.

2. The player then jumps, randomly and repeatedly, to any one of the four closest grid points from his current position. Since the resulting motion is random, the player is describing (in two dimensions) what is known as Brownian motion.

3. If the player lands on the outer conductor at any time, he is extracted from the grid and assigned a value of 0. The game is over.

4. If the player lands on the center conductor at any time, he is extracted from the grid and assigned a value of 1. The game is over.

Consider what happens when this game is repeated many times, each repetition starting from the same grid point. The ratio of the total number of 1's ("hits") to the total number of starts must approach the probability of being extracted at the center conductor.

Since while playing, the probability of a jump in any one direction is equal to the probability of a jump in any other direction, the probability at any time of reaching the center conductor is just the average of the probabilities of reaching the center conductor at each of the four points surrounding the player. In other words, if $P_{i,j}$ is the probability of reaching the center conductor from grid point

(i, j), then

$$P_{i,j} = \frac{P_{i+1,j} + P_{i-1,j} + P_{i,j+1} + P_{i,j-1}}{4} \tag{9.50}$$

By comparison, let us reexamine (9.20),

$$V_{i,j} = \frac{V_{i+1,j} + V_{i-1,j} + V_{i,j+1} + V_{i,j-1}}{4} \tag{9.20}$$

The obvious similarity between (9.50) and (9.20) suggests the following analogy: the probability of the player starting at any point on the grid and reaching the center conductor is identically the potential of the same starting grid point, assuming only that the center conductor is at 1 volt and the outer conductor is at 0 volts.

The computer algorithm for calculating the required probability requires the use of a random number generator. Enough players are started at each grid point to establish some measure of confidence in the resulting probability. This procedure is repeated at every grid point of interest. If these points are a set of grid points surrounding the center conductor, the required data for a total charge calculation are being acquired.

The computational advantages of using probabilistic potential theory rather than a very fine grid relaxation calculation for total charge data are twofold. First, the potential is being calculated only at points of interest, not over the entire grid. This saves computation time. Second, the only grid voltages that are saved are those of interest. This often reduces an overwhelming requirement for array storage to a trivial one.

Equation (9.50) and its analogous equation (9.20) are useful for uniform dielectric systems only. In the case of dielectric interfaces it is necessary to extend the analogy to assign probabilities of crossing a dielectric interface to agree with (9.38) and (9.49).

The example chosen is again a repetition of the single enclosed microstrip line problem (Figure 80). This allows us to compare the results to the previous results. As the listing for this example (Figure 86) suggests, there is very little algebraic calculation in the program. On the other hand, there is a very large amount of conditional branching. This is necessary because the player must choose a direction of jump based on the results of a random number generator, then check the location for the crossing of a dielectric interface or a hit on the center or outer conductor. The random number generator RANDU, referred to in the program, is standard to the mathematical subroutine libraries of most FORTRAN IV systems. It generates a random number between 0 and 1, starting with an initial (supplied) seed.

Another interesting feature of the program is the complete lack of array storage. This is in contrast to the relaxation grid solution of the same problem,

FORTRAN IV-PLUS V02-51D   22:13:10   08-DEC-78   PAGE 1
EX7.FTN   /TR:ALL/WR

```
         C  RANDOM WALK MICROSTRIP LINE EXAMPLE
         C
         C  INITIALIZATION
         C
         C  INITIALIZE RANDU
0001            DO 1  I = 1,100
0002          1 CALL RANDU(I1,I2,XOUT)
         C  SET DIELECTRICS
0003            ER = 10.
0004            E3 = (1. + ER)/2.
         C  TABULATE DIELECTRIC EDGE PROBABILITIES
0005            DENOM = 1. + ER + 2.*E3
0006            PDOWN = ER/DENOM
0007            PUP = 1./DENOM
0008            PSIDE = E3/DENOM
         C  SET DIMENSIONS ON BOX
0009            YLOW = 0.
0010            YHIGH = 50.
0011            XLOW = 0.
0012            XHIGH = 100.
         C  SET DIMENSIONS AND LOCATION OF CENTER CONDUCTOR
0013            IXCLOW = 46
0014            IXCHI = 55
0015            IYCLOW = 11
0016            IYCHI = 12
0017            XCLOW = IXCLOW
0018            XCHI = IXCHI
0019            YCLOW = IYCLOW
0020            YCHI = IYCHI
         C
0021            TYPE *, ' ', ' I        J       V', '
         C  PICK THE POINT OF INTEREST
0022            DO 500  I = IXCLOW - 1, IXCHI + 1
0023            DO 500  J = IYCLOW - 1, IYCHI + 1
0024            IF (I - (IXCLOW -1))  500, 9, 2
0025          2 IF (I - (IXCHI + 1))   3, 9, 500
0026          3 IF (J - (IYCLOW - 1)) 500, 9, 4
0027          4 IF (J - (IYCHI + 1)) 500, 9, 500
         C  ZERO THE COUNTERS
0028          9 NTOT = 0
0029            NHIT = 0
         C  SET STARTING POINTS
0030         10 X = I
0031            Y = J
         C  INCREMENT TOTALS COUNTER
0032            NTOT = NTOT + 1
         C  TAKE A RANDOM STEP
0033         20 CALL RANDU(I1, I2, XOUT)
         C  ARE WE ON THE DIELECTRIC EDGE?
0034            IF (Y .EQ. 11)  GO TO 40
         C  NO
0035            IF (XOUT - .25)  32, 32, 33
0036         32 X = X - 1.
0037            GO TO 60
0038         33 IF (XOUT - .5)  34, 34, 35
0039         34 X = X + 1.
```

(a)

Figure 86 Listing of Monte Carlo microstrip program.

to the same resolution, which required a 5000 element voltage array. Although 5000 elements is not a large number for most modern computer systems, it is conceivable that the resolution requirements could be 100 times smaller than in this example. A uniform relaxation grid would then require 50×10^6 elements, while the probabilistic potential theory program would still require zero.

Figure 87 shows the voltage grid values about the center conductor, super-

```
FORTRAN IV-PLUS V02-51D            22:13:10    08-DEC-78              PAGE 2
EX7.FTN              /TR:ALL/WR

0040                    GO TO 60
0041          35    IF (XOUT - .75)  36, 36, 37
0042          36    Y = Y - 1.
0043                    GO TO 60
0044          37    Y = Y + 1.
0045                    GO TO 60
         C   ON THE DIELECTRIC EDGE
0046          40    IF (XOUT - PDOWN)  42, 42, 43
0047          42    Y = Y - 1.
0048                    GO TO 60
0049          43    IF (XOUT - PDOWN - PUP)  44, 44, 45
0050          44    Y = Y + 1.
0051                    GO TO 60
0052          45    IF (XOUT - PDOWN - PUP - PSIDE)  46, 46, 47
0053          46    X = X - 1.
0054                    GO TO 60
0055          47    X = X + 1.
         C   CHECK FOR EXTRACTION
0056          60    IF ((X .LE. XLOW) .OR. (X .GE. XHIGH) .OR.
              2       (Y .LE. YLOW) .OR. (Y .GE. YHIGH ) GO TO 80
0057                IF ((X .GE. XCLOW) .AND. (X .LE. XCHI) .AND.
              2       (Y .GE. YCLOW) .AND. (Y .LE. YCHI)) GO TO 70
         C   KEEP MOVING
0058                    GO TO 20
         C   EXTRACT AND CHECK FOR ITERATION LIMIT - SET TO 200 FOR NOW
0059          70    NHIT = NHIT + 1
0060          80    IF (NTOT .LT. 200) GO TO 10
         C   CALCULATE THE VOLTAGE
0061                V = FLOAT(NHIT)/FLOAT(NTOT)
0062                TYPE 1000, I, J, V
0063        1000    FORMAT (2(I4,4X),F8.2)
0064         500    CONTINUE
0065                CALL EXIT
0066                END
```

(b)

Figure 86 *(Continued)*

imposed on the results of a 250 iteration relaxation grid solution (using over-relaxation) of the same problem. For 200 starts from each point, the probabilistic potential theory solution shows a large amount of scatter in results about the "true" values. However following the procedure for calculating Q, outlined in the preceding section, there is an integration involved. If the probabilistic potential theory solution can be thought of as, in some sense, the correct solution buried in random noise, then integrating around the circumference of the center conductor should tend to cancel the effects of the scatter on the total charge.

Repeating exactly the calculation procedure of the preceding section, the value of C is found to be $C = 192 \times 10^{-12}$. This is a slightly more accurate result than the total charge calculation based on the relaxation grid approach. The comments of the preceding section about going to a finer grid to increase the accuracy in the corner region apply equally well in this situation. Here, however, it is a simple matter to examine more closely spaced points—either around the entire circumference or just in the corner regions. Note, of course. that the numerical integration must be set up properly in terms of the space between the points.

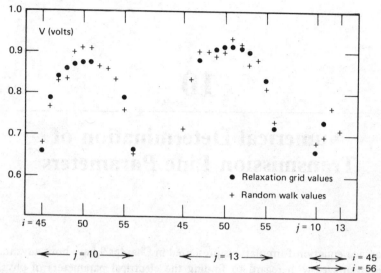

Figure 87 Output of Monte Carlo program (pluses) superimposed on output of relaxation grid program (circles).

To calculate the capacitances of multiline structures using probabilistic potential theory, it is necessary to generalize (9.50) to include several lines at arbitrary voltages. This generalization is

$$V_p = \frac{\sum_i V_i \text{(number of points extracted at } i\text{th electrode)}}{\sum_i \text{(number of points extracted at } i\text{th electrode)}} \qquad (9.51)$$

where i is summed over all electrodes present. In the case of nonidentical coupled lines, (9.51) will still apply.

9.11 SUGGESTED FURTHER READING

1. B. Carnahan et al., *Applied Numerical Methods*, Wiley, New York, 1969. Gives the numerical justifications of the convergence of the Gauss-Seidel form of the solution to Laplace's equation, along with a discussion of over-relaxation. Also derives procedures for handling unusual boundary conditions.

2. R. Geulusee, "Probabilistic Potential Theory Applied to Electrical Engineering Problems," *Proceedings of the IEEE*, Vol. 61, No. 4, April 1973.

10

Numerical Determination of Transmission Line Parameters

The Laplace equation formulations presented in Chapter 9 have both advantages and disadvantages with regard to finding the electrical parameters of physical transmission line systems. The principal advantage is conceptual simplicity. The numerical form of Laplace's equation is expressed in one simple algebraic equation, which is used repeatedly until the desired solution is obtained. One disadvantage is numerical inefficiency. Despite various convergence acceleration schemes, obtaining an accurate solution by relaxation methods is basically a war of attrition. Very high resolution in a problem complicated by different layers of dielectrics could challenge any computer budget. A second disadvantage is the premise of dealing only with problems having uniform cross sections. Thus far no technique has been proposed for studying abrupt discontinuities in conductor and/or dielectric cross sections of lines.

If the enclosed region described in Chapter 9 is considered to include the center conductor, then Laplace's equation must be replaced with Poisson's equation to properly take into account the charge on the center conductor. Poisson's equation can be solved by the method of Green's functions, one of the most elegant tools available in the field of mathematical physics. The Green's function for a particular geometry is often written as an infinite sum over some orthogonal set of basis functions. In the case of stripline and microstrip configurations, which are usually rectangular in cross section, the basis functions are sinusoids, and not too many terms need be considered to achieve very accurate results.

In a similar vein, it is often possible to express solutions of Laplace's equation —after a separation of variables—in terms of sums of sinusoids. Different sums must be used in different regions, and coefficients matched at the interfaces. The procedure is straightforward, and the results in many cases are more useful than the discrete set of numbers obtained from a relaxation grid.

Both the above-mentioned procedures yield, along with line parameter values, continuous function approximations for the charge densities on the different conductors. These approximations are very useful for studying current distributions in the lines, and in understanding which facets of a particular line geometry are critical in terms of line losses.

Since certain problems in transmission line networks are not two dimensional, they are not amenable to any of the approximation procedures discussed thus far. These problems include sudden changes in line widths, gaps between lengths of line, corners in lines, and holes in lines.

One approach to the three-dimensional problem is to subdivide a conductor into small rectangular "cells" and assume that the charge density is uniform in each cell. The voltage at the center of any one cell due to the charge density on any other cell is easily calculated; then the capacitance of the conductor is found by writing a "capacitance matrix" equation and solving it. The procedure, though a bit cumbersome, is straightforward and reasonably accurate.

10.1 POISSON'S EQUATION AND A GREEN'S FUNCTION SOLUTION

The Laplace's equation formulation of the line cross section problem treats the dielectric region as a finite area enclosed by boundaries that are held at fixed potentials. This (Laplace's equation) approach is valid only in charge-free regions. The problem can be treated more generally by considering the enclosed region as a layered dielectric region, with a charge density existing in some parts. In other words, we are now studying solutions to Poisson's equation. In two-dimensional rectangular coordinates, this is

$$\frac{\partial^2 V}{\partial x^2} + \frac{\partial^2 V}{\partial y^2} = -\frac{\rho(x, y)}{\epsilon} \tag{10.1}$$

where ρ is the charge density in the region.

As a practical case, consider a rectangular box with an infinitely thin conductor lying horizontal at some height in the box (Figure 88). This is, of course, a stripline. If two different dielectric regions,

$$\epsilon_r = \epsilon_{r1} \quad y < h$$
$$\epsilon_r = \epsilon_{r2} \quad y \geq h$$

are allowed, the problem is extended to include the quasi-static boxed microstrip. Also if a and b are allowed to get very large compared to h and w, the open microstrip is approximated very well.

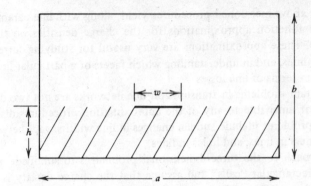

Figure 88　Stripline example for Green's function solution.

The applicable electrostatic boundary conditions are as follows:

$$V = 0 \quad \text{on the outer box} \tag{10.2a}$$

E_x is continuous at $y = h$; that is,

$$\frac{\partial V}{\partial x}(x, h-) = \frac{\partial V}{\partial x}(x, h+) \tag{10.2b}$$

D_y is continuous at $y = h$; that is,

$$\epsilon_1 \frac{\partial V}{\partial y}(x, h-) = \epsilon_2 \frac{\partial V}{\partial y}(x, h+) - \rho(x) \tag{10.2c}$$

If $\rho(x)$ were known, (10.1) could be solved directly in terms of the Green's function for the box:

$$V(x, y) = \int_{x_0} \rho(x_0)\, G(x, y, x_0)\, dx_0 \tag{10.3}$$

where $G(x, y, x_0)$ is the Green's function. This can be interpreted as the response in this particular geometry to a unit charge placed at (x_0, h). In other words, G is the solution of

$$\frac{\partial^2 G}{\partial x^2} + \frac{\partial^2 G}{\partial y^2} = \frac{-1}{\epsilon} \delta(x - x_0)\, \delta(y - h) \tag{10.4}$$

for the geometry of Figure 88.

Unfortunately, ρ is not known. If it were, the procedure above would be academic in terms of finding transmission line parameters. It would only be necessary to find Q by integrating ρ over the center conductor, and then $C = Q/V$. It is known, however, that a total charge Q, placed on the center conductor, will distribute itself so that the stored energy in the system is minimized. Furthermore, Thomson's theorem for electrostatics proves a point that is usually taken for granted: that is, the minimum energy charge distribution is identically the charge distribution that causes the voltage on the conducting strip to be uniform.

Implicit in the foregoing paragraph are several procedures for optimizing the parameters of an assumed form for ρ. Let us defer expansion of this statement until the function G has been found.

Assume that G can be written as the product of two functions, expanded in a Fourier series:

$$G = \sum_{n=1,3,5}^{\infty} f_n(y) \cos\left(\frac{n\pi x}{a}\right) \tag{10.5}$$

This equation satisfies the boundary condition

$$G\left(\frac{-a}{2}, y\right) = G\left(\frac{a}{2}, y\right) = 0$$

Substituting (10.5) into (10.4),

$$\sum_{1,3,5}^{\infty} \left[-\left(\frac{n\pi}{a}\right)^2 f_n(y) + \frac{d^2 f_n(y)}{dy^2} \right] \cos\left(\frac{n\pi x}{a}\right) = \frac{-1}{\epsilon} \delta(x \; x_0)\delta(y - y_0) \tag{10.6}$$

Following the procedure for finding Fourier series coefficients, multiply both sides of (10.6) by $\cos(m\pi x/a)$ and integrate across the box. Noting that the order of summation and integration may be interchanged, we have

$$\sum \left[-\left(\frac{n\pi}{a}\right)^2 f_n(y) + \frac{d^2 f_n}{dy^2} \right] \int_{-a/2}^{a/2} \cos\left(\frac{n\pi x}{a}\right) \cos\left(\frac{m\pi x}{a}\right) dx$$

$$= \frac{-1}{\epsilon} \int_{-a/2}^{a/2} \cos\left(\frac{m\pi x}{a}\right) \delta(x - x_0)\delta(y - y_0) \, dx \tag{10.7}$$

or

$$\left[-\left(\frac{n\pi}{a}\right)^2 f_n(y) + \frac{d^2 f_n(y)}{dy^2} \right] \frac{a}{2} = \frac{-1}{\epsilon} \cos\left(\frac{n\pi x_0}{a}\right) \delta(y - y_0) \tag{10.8}$$

The summation over n in (10.7) has disappeared because of the orthogonality properties of the integral.

For $y \neq y_0$, the right-hand side of (10.8) is zero. Remembering the boundary conditions at the top and bottom of the box, we write the solution to (10.8) as

$$f_n(y) = C_1 \sinh\left(\frac{n\pi y}{a}\right) \qquad y < h \qquad (10.9a)$$

$$f_n(y) = C_2 \sinh\left(\frac{n\pi}{a}(b - y)\right) \qquad y > h \qquad (10.9b)$$

In this problem, all charge resides at $y_0 = h$. Integrating (10.8) from $h-$ to $h+$ (an infinitesimally small step centered at h),

$$\frac{a}{2}\left[\epsilon \frac{df_n}{dy}\bigg|_{h_+} - \epsilon \frac{df_n}{dy}\bigg|_{h_-}\right] = -\cos\left(\frac{n\pi x_0}{a}\right) \qquad (10.10)$$

or

$$\frac{n\pi}{2}\left[\epsilon_2 C_2 \cosh\frac{n\pi}{a}(b - h) + \epsilon_1 C_1 \cosh\frac{n\pi h}{a}\right] = \cos\left(\frac{n\pi x_0}{a}\right) \qquad (10.11)$$

Equation 10.11 is a statement of (10.2c). The relation still needed comes from (10.2b): a relation between C_1 and C_2, which is found by equating (10.9a) and (10.9b) at $y = h$:

$$C_1 \sinh\left(\frac{n\pi h}{a}\right) = C_2 \sinh\left[\frac{n\pi}{a}(b - h)\right] \qquad (10.12)$$

Equation 10.11 is now

$$C_1\left[\epsilon_1 \cosh\left(\frac{n\pi h}{a}\right) + \epsilon_2 \frac{\sinh\left(\frac{n\pi h}{a}\right)\cosh\left(\frac{n\pi}{a}(b - h)\right)}{\sinh\left(\frac{n\pi}{a}(b - h)\right)}\right]$$

$$= \frac{2}{n\pi}\cos\left(\frac{n\pi x_0}{a}\right) \qquad (10.13)$$

and therefore

$$C_1 = \frac{\dfrac{2}{n\pi}\cos\left(\dfrac{n\pi x_0}{a}\right)\sinh\left(\dfrac{n\pi}{a}(b - h)\right)}{\epsilon_1 \cosh\left(\dfrac{n\pi h}{a}\right)\sinh\left(\dfrac{n\pi}{a}(b - h)\right) + \epsilon_2 \sinh\left(\dfrac{n\pi h}{a}\right)\cosh\left(\dfrac{n\pi}{a}(b - h)\right)}$$

$$(10.14)$$

$$C_2 = \frac{\dfrac{2}{n\pi} \cos\left(\dfrac{n\pi x_0}{a}\right) \sinh\left(\dfrac{n\pi h}{a}\right)}{\epsilon_1 \cosh\left(\dfrac{n\pi h}{a}\right) \sinh\left(\dfrac{n\pi}{a}(b-h)\right) + \epsilon_2 \sinh\left(\dfrac{n\pi h}{a}\right) \cosh\left(\dfrac{n\pi}{a}(b-h)\right)}$$

$$(10.15)$$

Finally, gathering together all the pieces,

$$G = \begin{cases} \displaystyle\sum_{n=1,3,5}^{\infty} C_1 \sinh\left(\frac{n\pi y}{a}\right) \cos\left(\frac{n\pi x}{a}\right) & y \leqslant h \\ \displaystyle\sum_{n=1,3,5}^{\infty} C_2 \sinh\left(\frac{n\pi}{a}(b-y)\right) \cos\left(\frac{n\pi x}{a}\right) & y \geqslant h \end{cases} \qquad (10.16)$$

Returning to the problem of finding C, a simple procedure can be derived based on Thomson's theorem. Along the center conductor,

$$V_0 = \int_{x_0} \rho(x_0') \, G(x, h, x_0', h) \, dx_0' \qquad (10.17)$$

Multiplying (10.17) by $\rho(x)$, we have

$$V_0 \rho(x) = \int_{x_0} \rho(x) \rho(x_0') \, G(x, h, x_0', h) \, dx_0' \qquad (10.18)$$

Now

$$C = \frac{Q}{V} = \frac{Q^2}{QV_0} = \frac{\left[\displaystyle\int_{x_0} \rho(x)\,dx\right]^2}{\displaystyle\int_{x_0} \int_{x_0'} \rho(x)\rho(x_0'') \, G(x, h, x_0'', h) \, dx_0'' \, dx} \qquad (10.19)$$

For a given total charge Q, ρ will arrange itself to minimize U. This in turn will maximize C in (10.19). This procedure will not predict ρ, but it will allow a calculation of C for any trial function for ρ. The parameters in the trial function must then be adjusted to maximize the resulting C. It can be shown that the value of C calculated by this procedure will always be less than or equal to the actual solution. The trial function that predicts the largest C is therefore the most accurate of those chosen. The form of (10.19) is convenient in that scale factors cancel, and only the functional form of ρ is considered.

As an example, consider a centered conductor of width w as shown in Figure 88. Let

$$\rho = \frac{1}{w}\left[1 + \frac{b}{w}|x|\right] \qquad 0 \leqslant |x| \leqslant \frac{w}{2} \tag{10.20}$$

Although this choice of ρ does not appear to be particularly physical (it has corners), it does demonstrate the basic attributes ρ should exhibit. Also, it is mathematically simple enough for a reasonable example. Substituting (10.20) into (10.19),

$$C = \frac{\left[\dfrac{1}{w}\displaystyle\int_0^{w/2}\left(1 + \dfrac{Bx}{w}\right)dx\right]^2}{\displaystyle\sum_{1,3,5}^{\infty} g_n(h)\left[\dfrac{1}{w}\displaystyle\int_0^{w/2}\left(1 + \dfrac{Bx}{w}\right)\cos\left(\dfrac{n\pi x}{a}\right)dx\right]} \tag{10.21}$$

where

$$g_n(h) = \frac{2}{n\pi}\,\frac{\sinh\left(\dfrac{n\pi h}{a}\right)\sinh\left(\dfrac{n\pi}{a}(b-h)\right)}{\epsilon_1\cosh\left(\dfrac{n\pi h}{a}\right)\sinh\left(\dfrac{n\pi}{a}(b-h)\right) + \epsilon_2\sinh\left(\dfrac{n\pi h}{a}\right)\cosh\left(\dfrac{n\pi}{a}(b-h)\right)} \tag{10.22}$$

Evaluating the integrals in (10.21),

$$C = \frac{\frac{1}{4}\left(1 + \dfrac{B}{4}\right)^2}{\displaystyle\sum g_n(h)\left[\dfrac{a}{wn\pi}\sin\left(\dfrac{n\pi w}{2a}\right) + B\left\{\dfrac{a}{2wn\pi}\sin\left(\dfrac{n\pi w}{2a}\right)\left(\dfrac{a}{wn\pi}\right)^2\left(\cos\left(\dfrac{n\pi w}{2a}\right) - 1\right)\right\}\right]^2}$$

$$= \frac{\frac{1}{4}\left(1 + \dfrac{B}{4}\right)^2}{\displaystyle\sum_n g_n(d_1 + Bd_2)^2} \tag{10.23}$$

where d_1 and d_2 are functions of N.

To find B, we maximize C. That is, set $dC/dB = 0$:

$$\sum g_n(d_1 + Bd_2)^2\,\frac{1}{4}\left(1 + \frac{B}{4}\right) = \left(1 + \frac{B}{4}\right)^2\sum g_n(d_1 + Bd_2)\,d_2 \tag{10.24}$$

Solving (10.24) for B,

$$B = \frac{-\sum g_n d_1 (d_1 - 4d_2)}{\sum g_n d_2 (d_1 - 4d_2)} \tag{10.25}$$

A computer program using these results to calculate C can conveniently also find several other variables of interest. Once ρ is known, the reasonability of the functional form of ρ can be examined by calculating the voltage along the center conductor $V(x, h)$:

$$V(x, h) = \int_{-w/2}^{w/2} \rho(x_0) \sum g_n(h) \cos\left(\frac{n\pi x}{a}\right) \cos\left(\frac{n\pi x_0}{a}\right) dx_0 \tag{10.26}$$

In terms of (10.20), the foregoing becomes

$$\frac{2}{w} \int_0^{w/2} \left(1 + \frac{Bx}{w}\right) \sum_n g_n(h) \cos\left(\frac{n\pi x}{a}\right) \cos\left(\frac{n\pi x_0}{a}\right) dx_0 \tag{10.27}$$

Another useful calculation is that of finding the return current density profiles along the outer conductor of the line (the outer box). The argument here is that the current flowing will be proportional to the charge profile, which in turn is proportional to ∇V. Now,

$$G = \begin{cases} \sum C_1 \sinh\left(\frac{n\pi y}{a}\right) \cos\left(\frac{n\pi x}{a}\right) & y \leqslant h \quad (10.28a) \\ \\ \sum C_2 \sinh\left(\frac{n\pi}{a}(b - y)\right) \cos\left(\frac{n\pi x}{a}\right) & y \geqslant h \quad (10.28b) \end{cases}$$

Therefore, along the bottom of the box,

$$D_y = -\epsilon_1 \int \rho(x_0) \sum C_1 \left(\frac{n\pi}{a}\right) \cos\left(\frac{n\pi x}{a}\right) dx_0 \tag{10.29}$$

and along the top of the box,

$$D_y = +\epsilon_2 \int \rho(x_0) \sum C_2 \left(\frac{n\pi}{a}\right) \cos\left(\frac{n\pi x}{a}\right) dx_0 \tag{10.30}$$

Typical results for these calculations appear in Figures 89 and 90. As can be seen, in a microstrip line most of the current in the outer conductor flows directly under the center conductor. This must be kept in mind when designing circuits that interconnect to stripline and microstriplines in various configurations. An

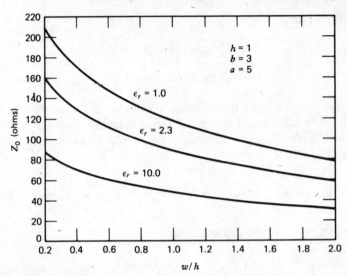

Figure 89 Capacitance calculated for the stripline by the Green's function method.

arbitrarily chosen point on the outer conductor is not ground in the sense that anything connected to that point is automatically at the same potential as every other point in the same line cross section. To achieve a minimum disturbance connection, the point of connection should be directly under the center conductor. On the other hand, the current in the upper conductor flows principally along the edges of the line. This is why loss calculations for microstrip and strip-

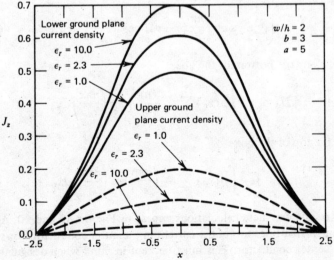

Figure 90 Current densities in the ground planes, calculated by the Green's function method.

line based on the incremental inductance rule always predict lower losses for thicker lines—well past the thickness limit that would be predicted by a simple calculation of several skin depths of thickness: the current is flowing mainly along the edges, and the edges must be as "thick" as practical.

The accuracy of the results obtained for C are surprisingly good for the simple form of p chosen. This accuracy can be improved by using a more sophisticated trial function for ρ. Many terms of a power series in x can be used, for example. The tradeoffs here are those of computational difficulty: a many-termed approximation also has many unknown coefficients, and the process of maximizing C becomes quite involved. The maximum value of C could also be found by numerical search techniques, but in this case analytic derivation effort is being traded for costly computer search effort. Before a decision can be made, in any case, individual requirements must be scrutinized in terms of preparation time, computer time, intended use, and required accuracy.

10.2 AN ANALYTIC LAPLACE'S EQUATION SOLUTION FOR STRIPLINE

The possibilities for an analytic solution to the problem of finding the capacitance of a transmission line increase greatly with the symmetries and resultant simplifications possible. The best example of this is, of course, the coaxial line. The coaxial line can be solved exactly in a very simple manner. Similarly, a parallel plate line is simple to approximate if fringing capacitances may be ignored— that is, if the width-to-separation ratio of the plates is large.

The homogeneous stripline problem can be approximated quite well under certain conditions by "piecing together" a solution using various techniques. Figure 91 shows the cross section of a homogeneous rectangular stripline in a closed box. The center conductor is assumed to have zero thickness, although the procedure to be described will work equally well for a finite thickness center conductor—with only a moderate amount of additional calculation necessary.

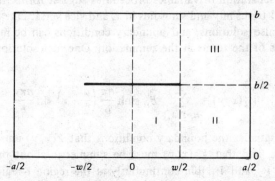

Figure 91 Stripline example for the analytic Laplace's equation solution example.

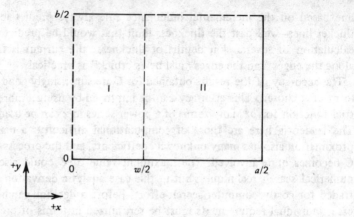

Figure 92 Subsection of the stripline example, showing choice of axes.

As Figure 91 indicates, the line is symmetric in the four quadrants taken about the center. This means that the capacitance of each quadrant is the same as the others and equal to one-quarter of the total. It is therefore necessary to consider only one quadrant. Each quadrant can be divided further into two regions—the region under the center conductor and the region not under the center conductor. These regions are labeled I and II, respectively, for the lower right-hand quadrant of the line in Figure 91.

Figure 92 shows the lower right-hand quadrant of the line alone with the choice of axes and relevent dimensions labeled. If $w \gg b$, the voltage in region I should be approximately

$$V_1(x,y) = \frac{2V_0 y}{b} \tag{10.31}$$

The voltage in region II must satisfy the necessary boundary conditions and connect continuously with the approximate solution for the voltage in region I. Conventional separation of variables procedures suggest forms that are products of exponential terms in y and sinusoids in x, and vice versa. Linear sums of these solutions are also solutions, and boundary conditions can be met by adjusting the amplitudes of the terms in the summation. One such solution for V_2 would be

$$V_2(x,y) = \sum_{n=1,3,5}^{\infty} B_n \sinh \frac{n\pi}{b} \left(\frac{a}{2} - x\right) \sin \frac{n\pi y}{b} \tag{10.32}$$

This solution satisfies the boundary conditions that $V(x,y)$ must be zero along the outer box, and that $V_2(x,y)$ must be symmetric about the line $y = b/2$. In order that V_2 and V_1 join continuously at the region I–region II interface, we must have

$$V_2\left(\frac{w}{2}, y\right) = \sum_{1,3,5} B_n \sinh \frac{n\pi}{2b}(a - w) \sin \frac{n\pi y}{b} \equiv \sum_{1,3,5} C_n \sin \frac{n\pi y}{b} \equiv \frac{2V_0 y}{b}$$

$$(10.33)$$

The coefficients C_n can be found by conventional Fourier series procedures—that is, by taking advantage of the orthogonality properties of the sinusoids,

$$\int_{-b/2}^{b/2} \sin \frac{n\pi y}{b} \sum C_n \sin \frac{n\pi y}{b} dy = C_n \int_{-b/2}^{b/2} \sin^2 \frac{n\pi y}{b} dy$$

$$= \int_{-b/2}^{b/2} \frac{2V_0 y}{b} \sin \frac{n\pi y}{b} dy \quad (10.34)$$

Solving (10.34) for C_n,

$$C_n = \frac{8V_0}{n^2\pi^2} \sin \frac{n\pi}{2}$$

$$(10.35)$$

and therefore

$$B_n = \frac{8V_0 \sin\left(\frac{n\pi}{2}\right)}{n^2\pi^2 \sinh \frac{n\pi}{2b}(a - w)}$$

$$(10.36)$$

It should be noted that in general, Fourier series theory should not be applied to the problem of finding C_n in (10.33), since the absence of the even terms in the summation negates the assurance that a complete set of coefficients will be found to match the desired wave shape. On the other hand, the even numbered terms could not be included because they do not satisfy the boundary condition requirement for symmetry about the line $y = b/2$—that is, $\partial V_2/\partial y \big|_{y=b/2} \neq 0$. Fortunately, this problem resolves itself in this case, since C_n is identically zero when n is even, and the full guarantees of Fourier series theory apply.

The electric fields in region II are

$$E_x = \sum \frac{n\pi}{b} B_n \cosh \frac{n\pi}{b}\left(\frac{a}{2} - x\right) \sin \frac{n\pi y}{b}$$

$$(10.37)$$

$$E_y = -\sum \frac{n\pi}{b} B_n \sinh \frac{n\pi}{b}\left(\frac{a}{2} - x\right) \cos \frac{n\pi y}{b}$$

$$(10.38)$$

The energy stored in the electric field in region II is therefore

$$U_{II} = \frac{1}{2} \epsilon \int_0^{-b/2} \int_{w/2}^{a/2} \sum_n \sum_m mn \frac{\pi^2}{b^2} B_m B_n$$

$$\cdot \left[\sinh^2 \frac{m\pi}{b} \left(\frac{a}{2} - x\right) \sinh^2 \frac{n\pi}{b} \left(\frac{a}{2} - x\right) \cos \frac{m\pi y}{b} \cos \frac{n\pi y}{b} \right.$$

$$\left. + \cosh^2 \frac{m\pi}{b} \left(\frac{a}{2} - x\right) \cosh^2 \frac{n\pi}{b} \left(\frac{a}{2} - x\right) \sin \frac{m\pi y}{b} \sin \frac{n\pi y}{b} \right] dx\, dy$$

$$(10.39)$$

Again, the orthogonality properties of the sinusoids simplify the expressions greatly, and (10.39) becomes

$$U = \frac{\epsilon}{2} \int_{w/2}^{a/2} \sum_{n=1,3,\ldots}^{\infty} \frac{n^2 \pi^2 B_n^2}{4b} \left[\sinh^2 \frac{n\pi}{b} \left(\frac{a}{2} - x\right) + \cosh^2 \frac{n\pi}{b} \left(\frac{a}{2} - x\right) \right] dx$$

$$= \sum_{n=1,3,5}^{\infty} \frac{4\epsilon V_0^2}{n^3 \pi^3} \frac{\sinh \frac{n\pi}{b} (a - w)}{\sinh^2 \frac{n\pi}{2b} (a - w)}$$

$$(10.40)$$

In the full line cross section there are four regions (in total) identical to region II. Also, since $U = \frac{1}{2} CV^2$ for a capacitor, the total capacitance contribution of "all of the regions II" is 8 times (10.40). The parallel plate capacitance of the structure (region I, 4 times over) is simply $4w/b$; therefore the total capacitance per unit length of the enclosed stripline is

$$C_{tot} = \epsilon \left[\frac{4w}{b} + \frac{32}{\pi^3} \sum \frac{1}{n^3} \frac{\sinh \frac{n\pi}{b} (a - w)}{\sinh^2 \frac{n\pi}{2b} (a - w)} \right]$$

$$(10.41)$$

Figure 93 shows (10.41) plotted for $a = 1$, versus w for several values of b. Shown in dashed lines are the values predicted by an exact solution to the same problem (Reference 5, Chapter 6). Clearly (10.41) is quite accurate when b is much smaller than a, but poor when b is approximately equal to a. The reasons for these conclusions can be found by examining the physical nature of the approximations made at the beginning of this section. When $b \ll a$, the parallel plate capacitance dominates, and the fringing capacitance merely causes a small positive offset in the total capacitance. The electric field lines in region I will be

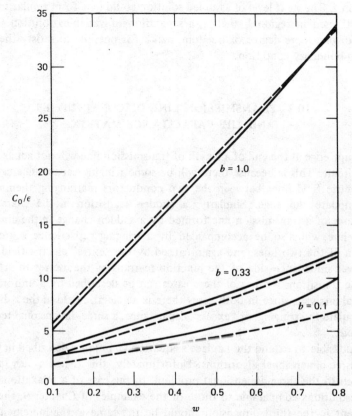

Figure 93 Results of analytic Laplace's equation stripline example compared to exact results.

essentially vertical except for a very small region near the region I–region II boundary. On the other hand, when $b \approx a$, the electric field lines under the center conductor (region I) will have a considerable x component, and the approximation for V_1 is poor. If it is necessary to derive a more accurate equation for the $b \approx a$ case, a more sophisticated assumed voltage form is necessary in region I. The foregoing derivation seemed to "fit" all the boundary conditions, yet was only an approximation, because one necessary boundary condition was ignored: not only must $V(x, y)$ be continuous across the region I–region II interface, but $\partial V/\partial x$ must also be continuous across the same interface. The form of the solution used above, in a sense, postulates a linear charge density at $x = w/2$ that mathematically satisfies Poisson's equation across the interface, thereby physically accounts for the totally vertical electric field lines in region I. The next level of assumed solution could be a form similar to that in region II used in region I—that is, a separation of variables solution that will allow for one more degree of freedom, making it possible to satisfy the one remaining boundary condition.

10.3   TRANSMISSION LINE DISCONTINUITIES AND THE CAPACITANCE MATRIX

An abrupt edge at the end of a length of transmission line will not act as an ideal open circuit. This is because there will be some fringing capacitance caused by the electric field lines between the two conductors rearranging themselves to accommodate the edge. Similarly, a sudden transition in the characteristic impedance of a transmission line formed by a sudden change in the dimensions of the line will also be accompanied by a fringing capacitance shunting the junction of the two lines, and again caused by the "extra" electric field lines of the lower impedance side of the junction rearranging themselves to accommodate the transition. Neither of these cases can be described by a uniform cross-sectional analysis, since in both cases there is an electric field in the z direction. To calculate the fringing, or "excess" capacitance, a three-dimensional technique is needed.

It is possible to extend the Laplace's equation techniques described in Chapter 9 to three-dimensional algorithms. Unfortunately, this is not a very practical approach to the three-dimensional problem. In the case of a relaxation grid, to hold resolutions comparable to those of the examples of Chapter 9, the spatial increment in the third dimension should be the same as the increment in the cross-sectional dimensions, and a typical problem would require a computer storage capability in the order of $100 \times 50 \times 50$, or 250,000 real variables. This is not realistic in terms of today's machine capabilities. Also, the number of iterations required for satisfactory accuracy would increase tremendously. In the case of a probabilistic potential theory solution, the Brownian motion is now in a three-dimensional box, and the expected time of particle "wandering" goes up in a manner similar to the expected relexation time of the grid.

The approach chosen to solve the problems described is known as the *capacitance matrix* approach. It is basically a three-dimensional technique that is reducible to a two-dimensional technique when only one of the conductors of a transmission line has finite dimensions (the microstrip or stripline or slotline, etc., situations). Unfortunately, the approach is not as straightforward as that of the Laplace's equation technique, but with careful layout of algorithms, efficient and accurate solutions can be had.

Consider first a planar conductor with a charge density $\sigma(x, y)$ on it. Now subdivide the conductor into squares, each of size $2b \times 2b$. The voltage at the center of any given square (x_p, y_p) due to the charge on any (possibly even the same) square is

$$V_p = \int_{-b}^{b} \int_{b}^{b} \frac{\sigma(x,y)\,dx\,dy}{4\pi\epsilon\sqrt{(x-x_p)^2+(y-y_p)^2+z^2}} \tag{10.42}$$

where, since only the relative locations of the two squares to each other is of interest, we are assuming that the charged square is centered at the origin of the (x,y,z) system.

If σ is a constant in (10.42),

$$V_p = \frac{\sigma}{4\pi\epsilon} \int_{-b}^{b} \int_{b}^{b} \frac{dx\,dy}{\sqrt{(x-x_p)^2+(y-y_p)^2+z^2}} \tag{10.43}$$

For convenience in the calculations that follow, let us transform the variables in (10.43) according to

$$x = bu \tag{10.44a}$$

$$y = bv \tag{10.44b}$$

yielding

$$V_p = \frac{\sigma b}{4\pi\epsilon} \int_{-1}^{1} \int_{-1}^{1} \frac{du\,dv}{\sqrt{(u-u_p)^2+(v-v_p)^2+z^2/b^2}} \tag{10.45}$$

The total charge on the (charged) square is

$$\alpha = \int_{-b}^{b} \int_{b}^{b} \sigma\,dx\,dy = 4\sigma b^2 \tag{10.46}$$

and therefore

$$V_p = \frac{\alpha}{16\pi\epsilon_0 b} \int_{-1}^{1} \int_{-1}^{1} \frac{du\,dv}{\sqrt{(u-u_p)^2+(v-v_p)^2+(z/b)^2}} \tag{10.47}$$

This integral can be written more conveniently in terms of the separation between squares, which may be written in terms of integers only. That is, letting

$$x' = u - u_p \tag{10.48a}$$

$$y' = v - v_p \tag{10.48b}$$

(10.47) becomes

$$V_p = \frac{\alpha}{16\pi\epsilon_0 b} \int_{2N_v-1}^{2N_v+1} \int_{2N_u-1}^{2N_u+1} \frac{dx'\,dy'}{\sqrt{x'^2+y'^2+(z/b)^2}} \tag{10.49}$$

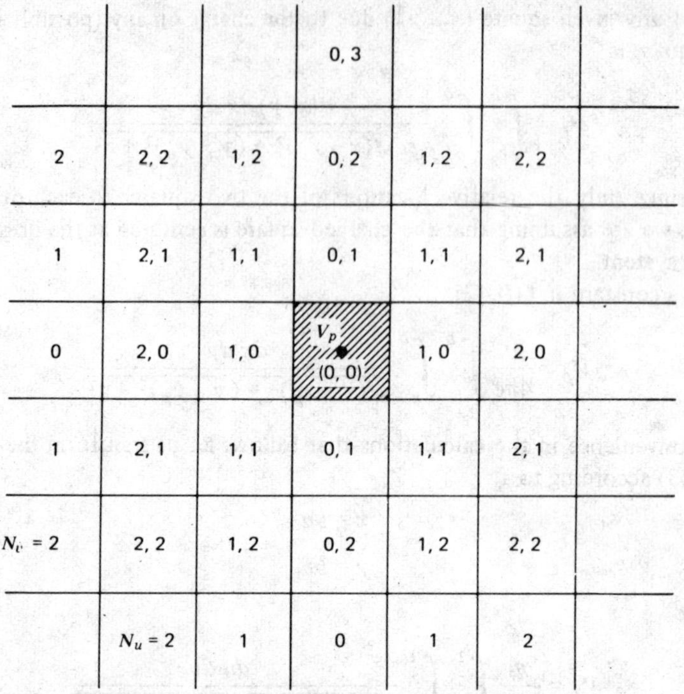

			0, 3		
2	2, 2	1, 2	0, 2	1, 2	2, 2
1	2, 1	1, 1	0, 1	1, 1	2, 1
0	2, 0	1, 0	V_p (0, 0)	1, 0	2, 0
1	2, 1	1, 1	0, 1	1, 1	2, 1
$N_v = 2$	2, 2	1, 2	0, 2	1, 2	2, 2
	$N_u = 2$	1	0	1	2

Figure 94 Format of incremental cell numbering system for cells in a plane.

where N_u and N_v are the number of cells between the source cell and the field cell in the x and y directions, respectively. Figure 94 gives several examples for an isoplanar structure. Note of course that the value of V_p is the same for an interchange of N_u and N_v—that is, $V_p(N_{u0}, N_{v0}) = V_p(N_{v0}, N_{u0})$.

Now, define $L_{i,j}$ as the voltage on the ith cell due to a charge of $\alpha = 1$ on the jth cell, and then (10.49) can be written as

$$V_{i,j} = L_{i,j}\alpha_j \qquad (10.50)$$

where

$$L_{i,j} = \frac{1}{16\pi\epsilon_0 b} \int_{2N_v-1}^{2N_v+1} \int_{2N_u-1}^{2N_u+1} \frac{dx'\, dy'}{\sqrt{x'^2 + y'^2 + (z/b)^2}} \qquad (10.51)$$

and $V_{i,j}$ is understood to be the voltage at the center of the ith cell.

Let us consider a structure made up of one or more planes, with 1 volt on one (or more) planes and 0 volts on zero or more planes—obviously 1 volt and 0 volts cannot be present together on the same conductor—then (10.50) can be generalized to the matrix equation

$$V = L\alpha \tag{10.52}$$

where the indices (i, j) extend over the cells on all the planes in the structure.

The total capacitance of the structure as described is simply the sum of the charge on all the cells that have 1 volt applied. That is,

$$C_{total} = \alpha \cdot V \tag{10.53}$$

The problem of finding the capacitance of an arbitrary (set of planes) structure can be divided into three principal subproblems:

1. Examining the given geometry and assigning cell locations.
2. Evaluating the coefficients $L_{i,j}$ for the given geometry.
3. Solving the matrix equation for the cell charges.

Item 1 is a matter of organization and bookkeeping. It is a job that must be done, and the efficiency of the resulting computer algorithm depends on its being done carefully. Unfortunately, it is totally specific to a given problem, and few generalizations can be made. Item 3 turns out to be a relatively straightforward procedure, because a simple Gauss-Seidel iteration converges very quickly.

Item 2 may or may not be a difficult job, as the examples below indicate. There are $4N_u^2 - 1 \times 4N_v^2 - 1$ entries in the $L_{i,j}$ matrix. Fortunately, many of these are identical to others, and far fewer than the full number of matrix positions must be evaluated. The integral (10.51) has been evaluated analytically. The resulting expression is lengthy, and if many thousand computer evaluations are necessary it is desirable to derive some approximate expressions for the integral:

Physically, if the source cell is far removed from the field point, it should not matter much if the charge is considered to be concentrated at the center of the source cell, at a point. In this case,

$$L = \frac{1}{8\pi\epsilon_0 b\sqrt{N_u^2 + N_v^2 + (z/2b)^2}} \tag{10.54}$$

Table 8 shows values of the exact integral and (10.54) for several values of N_u, N_v, and z/b. As (10.54) shows, is accurate to better than $1\frac{1}{2}\%$ except for the cells immediately adjacent to the source cell and for the cells directly "below" the source cell. In the former case, there are only two possible values of $L_{i,j}$ to consider—the $(1, 1)$ value and the $(1, 0) = (0, 1)$ value. It is a simple matter to note these values and use them when required. In the latter case—that is, $N_u = N_v = 0$ and $z/b \neq 0$—the following approximation is valid to better than $1\frac{1}{2}\%$ for

Table 8 $L_{i,j}$ Exact Values and Two Approximations

N_u	N_v	z/b	I_{exact}	$I_{point\ charge}$	$I_{approx\ form}$
0	0	0	0.1403	–	0.1403
0	1	0	0.0413	0.0398	–
1	1	0	0.0288	0.0281	–
0	2	0	0.0201	0.0199	–
0	0	1	0.0632	0.0796	0.0624
0	1	1	0.0359	0.0356	–
1	1	1	0.0269	0.0265	–
0	2	1	0.0195	0.0193	–
0	0	2	0.0370	0.0398	0.0373
Etc.					

all values of z/b:

$$L = \frac{1}{\epsilon_0 b} \left[\frac{1}{12.563\,(z/b) + 7.128 e^{-0.719 z/b}} \right] \qquad (10.55)$$

As a first example, consider the case of a single rectangular charged conductor, of dimensions 10×2 m². Let $b = 1$, so that there are 5 cells, each 2×2 m², in a row. Label them 1 to 5, in order, as in Figure 95. For convenience, let us drop the ϵ_0 in (10.51). Since the calculations are linear, we can get C by multiplying the resulting total charge by ϵ_0.

The L_{ij} matrix has 25 terms. However, there are only 5 different values. These are:

$$L_{11} = L_{22} = L_{33} = L_{44} = L_{55} = 0.1403$$

$$L_{12} = L_{21} = L_{23} = L_{32} = L_{34} = L_{43} = L_{45} = L_{54} = 0.0413$$

$$L_{13} = L_{31} = L_{24} = L_{42} = L_{35} = L_{53} = 0.0201$$

$$L_{14} = L_{41} = L_{25} = L_{52} = 0.0133$$

$$L_{15} = L_{51} = 0.0099$$

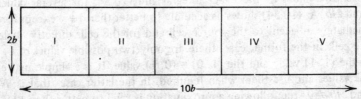

Figure 95   Simple rectangular conductor example, capacitance matrix technique.

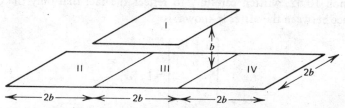

Figure 96 Simple three-dimensional parallel plane example, capacitance matrix technique.

The matrix equation (10.52) is further simplified by the symmetry of the problem. By inspection, $\alpha_1 = \alpha_5$ and $\alpha_2 = \alpha_4$. With or without taking advantage of this simplification, the solution of (10.52) for this example is

$$\alpha_1 = \alpha_5 = 4.85$$

$$\alpha_2 = \alpha_4 = 3.66$$

$$\alpha_3 = 3.58$$

The capacitance of the rectangular conductor is therefore

$$\epsilon_0(4.85 + 3.66 + 3.58 + 3.66 + 4.85) = 182 \text{ pF}$$

In an example of a three-dimensional problem (Figure 96), a conducting square plate is located centered over a conducting rectangular plate, with dimensions as shown. Letting $b = 1$ and $h = 1$, and numbering the cells as shown, leads to the following values for $L_{i,j}$:

$$L_{11} = L_{22} = L_{33} = L_{44} = 0.1403$$

$$L_{13} = L_{31} = 0.0632$$

$$L_{23} = L_{32} = L_{34} = L_{43} = 0.0413$$

$$L_{24} = L_{42} = 0.0201$$

$$L_{12} = L_{21} = L_{14} = L_{41} = 0.0359$$

In the first example the potential on the (single) conducting plate was referenced to a point at infinity. In this example there are two potentials present, and only the difference between them is preset to 1 volt—the individual values of the voltages is also a function of the total charge on the system. Assuming that we are interested only in the capacitance between the two conducting plates, we must add the constraint that

$$\sum_{i=1}^{\text{all cells}} \alpha_i = 0 \qquad (10.56)$$

Equation 10.52, written carefully to reflect the fact that only the potential difference between the plates is known, is

$$
\begin{bmatrix} 1 + V_r \\ V_r \\ V_r \\ V_r \end{bmatrix} = L_{i,j}\alpha_j
\tag{10.57}
$$

The symmetry of the structure requires that $\alpha_2 = \alpha_4$. Solving for the charges on the cells yields $\alpha_1 = 8.90$, $\alpha_2 = \alpha_4 = -2.60$, $\alpha_3 = -3.70$. To find the capacitance between the conductors, (10.53) must be modified to sum up either all the positive charge on one conductor or (equivalently) all the negative charge on the other conductor. In either case, $C = 78.8$ pF.

Both the examples above give indications of how charge distributes itself on transmission line cross sections. In the first example, the charge on a single conductor is seen to bunch at the outer edges of the conductor. This distribution, due to the self-repulsion of like charges, is directly analogous to the charge distribution on the upper conductor of microstrip or the center conductor of stripline. In the second example, the charge on a conductor that is wider than the nearby conductor carrying the negative of the same charge is seen to bunch as closely as possible to the nearby conductor. This is of course due to the attraction of opposite charges. This, in turn, is directly analogous to the charge distribution in the ground plane(s) in microstrip and stripline, where the charge distributes itself so as to establish equilibrium between its self-repulsion and opposite-attraction.

As a third example, consider a narrow strip of conductor centered between a pair of parallel plane conductors—in other words, an idealized stripline. By studying such a structure for several different length lines of the same width, it should be possible to separate the capacitance per unit length (the usual transmission line capacitance) and the end or fringing capacitance due to an abrupt edge. Also, by comparing the results for different width lines with the result obtained for an abrupt transition in line width, the "excess" capacitance of the transition should be obtainable.

To obtain reasonably accurate results, the cell structure must be fine enough to allow for physically realistic charge distributions. It becomes apparent at this point that to solve this problem by the methods used for the previous examples, an inordinately large number of cells must be used. An alternate approach in this case is to note that the problem can be reduced to that of a single plane conductor (the center strip) by considering the charge in the ground planes to be merely the image charge caused by the charge on the center strip.

Consider a single cell of uniformly distributed charge placed parallel to and centered between a pair of infinite parallel conducting planes. The planes are

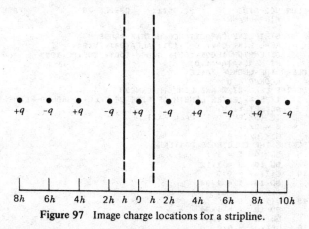

Figure 97 Image charge locations for a stripline.

separated by a distance $2h$. Figure 97 shows the resulting set of image charges: The charge distribution on the center strip ($+\alpha$) causes a pair of image charges ($-\alpha$) a distance $2h$ away from the center strip. Each of these images, in turn, causes an image ($+\alpha$) located at a distance $4h$ from the center strip. Each of these images, in turn, causes another pair of images, and so on, ad infinitum. Fortunately, the signs of the image charges alternate, and the resulting potential at any point centered between the ground planes is the sum of a converging series of terms. That is, if we extend the definition of $L_{i,j}$ to be the potential at any cell centered between the ground planes due to the charge on any other cell centered between the ground planes and all its images, (10.52) is still valid, and

$$L_{i,j} = \sum_{h=0,\pm2,\pm4\ldots}^{\infty} \frac{1}{16\pi\epsilon_0 b} \int_{2N_v-1}^{2N_v+1} \int_{2N_u-1}^{2N_u+1} \frac{dx\, dy}{\sqrt{x^2 + y^2 + (h/b)^2}} (-1)^{h/2}$$

$$(10.58)$$

The assertion that the series (10.58) converges can be seen by noting that after some number of terms, it must be reasonable to use the point source approximation for charge distribution (10.54); then (10.58) becomes

$$L_{i,j} \approx (h = 0 \text{ term}) + 2(h = 2 \text{ term}) + \cdots$$

$$+ \frac{2}{8\pi\epsilon_0 b} \left(\frac{1}{Nh} - \frac{1}{(N+1)h} + \frac{1}{(N+2)h} - \cdots \right) \quad (10.59)$$

Unfortunately (10.59) converges very slowly, and many terms are needed to assure accuracy.

The FORTRAN program presented in Figure 98 is a complete program for calculating the capacitance to ground of arbitrary center conductor geometries that can be described adequately using a 7×12 array of cells. The program is written

```
              C          STRIPLINE CAPACITANCE MATRIX PROGRAM
0001                     REAL L(84,84), C(7,12), ALFA(84), V(84)
0002                     DIMENSION ILOOK(100), JLOOK(100), COFTBL(100)
0003                     PI = 4.*ATAN(1.0)
              C  CLEAR THE LOOKUP TABLE
0004                     LOOKUP = 0
              C  DEFINE CELL SIZE AND LINE THICKNESS
0005          1          TYPE *,'ENTER LENGTH OF SIDE OF CELL,  GROUND PLANE SEPARATION'
0006                     ACCEPT *, B, H
0007                     IF (B .EQ. 0.)   CALL EXIT
0008                     B = B/2.
0009                     H = H/2./B
              C  DEFINE THE ELECTRODE PATTERN
0010          5          DO 10  I = 1,7
0011                     DO 10  J = 1,12
0012          10         C(I,J) = 0.0
0013                     DO 15  I = 1,84
0014                     ALFA(I) = 0.
0015          15         V(I) = 0
0016                     DO 18  I = 1,84
0017                     DO 18  J = 1,84
0018          18         L(I,J) = 0.0
0019          20         TYPE *, 'ELECTRODE: ENTER I,J - START; I,J - STOP, 0 TO CONT'
0020                     ACCEPT *, I1, J1, I2, J2
0021                     IF (I1 .EQ. 0.)  GO TO 50
0022                     DO 30  I = I1,I2
0023                     DO 30  J = J1,J2
0024          30         C(I,J) = 1.0
0025                     GO TO 20
              C  TRANSFER DATA TO V ARRAY
0026          50         DO 60  I = 1,7
0027                     DO 60  J = 1,12
0028                     K = (I-1)*12 + J
0029          60         V(K) = C(I,J)
              C  SET UP L ARRAY
              C  1ST GET CORRECT CELL
0030                     TYPE *,'     M       N      IM      JM      IN      JN      L
0031                     DO 200  M = 1,84
0032                     DO 200  N = 1,M
0033                     CALL LOCATE(M,N,IM,IN,JM,JN)
0034                     IF ((C(IM,JM) .EQ. 0.).OR.(C(IN,JN) .EQ. 0.)) GO TO 200
              C  FIND INCREMENTAL POSITION
0035                     IDELT = IABS(IM-IN)
0036                     JDELT = IABS(JM-JN)
              C  CHECK THE LOOKUP TABLE
0037                     IF (LOOKUP .EQ. 0)  GO TO 95
0038                     DO 90  I = 1,LOOKUP
0039                     IF (MINO(IDELT,JDELT) - ILOOK(I) )  90,75,90
0040          75         IF (MAXO(IDELT,JDELT) - JLOOK(I) )  90,80,90
0041          80         L(M,N) = COFTBL(I)
0042                     GO TO 200
0043          90         CONTINUE
0044          95         L(M,N) = COEF(IDELT,JDELT,0.)
0045                     DO 100  I = 2,15000,4
0046                     D = H*FLOAT(I)
0047                     L(M,N) = L(M,N) - 2.*( COEF(IDELT,JDELT,D) - COEF(IDELT,JDELT,D+2
```

(a)

Figure 98 FORTRAN program listing for a stripline with discontinuites, capacitance matrix technique.

```
0048      100    CONTINUE
0049             L(M,N) = L(M,N)/B
          C  LOAD AND UPDATE THE LOOKUP TABLE
0050             LOOKUP = LOOKUP + 1
          D      TYPE *, 'LOOKUP TABLE ENTRY NR', LOOKUP
0051             ILOOK(LOOKUP) = MINO(IDELT,JDELT)
0052             JLOOK(LOOKUP) = MAXO(IDELT,JDELT)
0053             COFTBL(LOOKUP) = L(M,N)
0054             IF (L(M,N) .NE. 0.)   TYPE 1000, M, N, IM,JM,IN,JN,L(M,N)
0055     1000    FORMAT (6I7,G12.4)
0056      200    CONTINUE
0057             DO 250  M = 1,84
0058             DO 250  N = M,84
0059      250    L(M,N) = L(N,M)
          C  INVERT THIS MESS
0060             CALL SOLVER(V,L,ALFA)
          C  GET TOTAL CAPACITANCE
0061             CAP = 0.
0062             DO 300  I = 1,84
0063      300    CAP = CAP + ALFA(I)
0064             CAP = CAP*8.854
0065             TYPE 1100, CAP
0066     1100    FORMAT (' CAPACITANCE TO GROUND = ',G10.3)
          C  GET CHARGE DISTRIBUTION
0067             TYPE *, 'ENTER 1 FOR CHARGE DISTRIBUTION'
0068             ACCEPT *, IOPT
0069             IF (IOPT)  320, 5, 340
0070      320    CALL EXIT
0071      340    TYPE *, '    NR        I        J        Q'
0072             DO 400  I = 1,84
0073             IX = (I - 1)/12 + 1
0074             JY = I - 12*(IX - 1)
0075             C(IX,JY) = ALFA(I)*8.854
0076      400    IF (ALFA(I) .NE. 0.)  TYPE 1200, I,IX,JY,C(IX,JY)
0077     1200    FORMAT (3I7,G12.3)
0078             TYPE *, ' ', ' '
0079             GO TO 5
0080             END
```

PROGRAM SECTIONS

NAME	SIZE		ATTRIBUTES
$CODE1	003376	895	RW,I,CON,LCL
$PDATA	000350	116	RW,D,CON,LCL
$IDATA	000146	51	RW,D,CON,LCL
$VARS	072612	15045	RW,D,CON,LCL
$TEMPS	000012	5	RW,D,CON,LCL

TOTAL SPACE ALLOCATED = 076740 16112

(b)

Figure 98 (*Continued*)

```
      C  SIMULTANEOUS EQUATION SOLVER
0001          SUBROUTINE SOLVER(V, L, ALFA )
0002          REAL V(84), L(84,84), ALFA(84), LNEW
      C  TRY A GAUSS-SEIDEL ITERATION
      C  START TO LOOP
0003          DO 400  NRPASS = 1,250
0004          ERRSUM = 0.
0005          ERRMAX = 0.
0006          DO 350  I = 1,84
0007          IF (L(I,I) .EQ. 0.)  GO TO 350
0008          LNEW = 1.0
0009          DO 300  J = 1,84
0010          IF (J .EQ. I)  GO TO 300
0011          LNEW = LNEW - ALFA(J)*L(I,J)
0012    300   CONTINUE
0013          LNEW = LNEW/L(I,I)
      C       TYPE *, I, J, LNEW, ALFA(I)
0014          ERR = ABS(LNEW - ALFA(I))**2
0015          ERRSUM = ERRSUM + ERR
0016          IF (ERR .GT. ERRMAX)  ERRMAX = ERR
0017          ALFA(I) = LNEW
0018    350   CONTINUE
      D       TYPE *, ERRMAX, ERRSUM
0019          IF (ERRSUM .LT. 1.E-8)  RETURN
0020    400   CONTINUE
0021          RETURN
0022          END
```

```
      C  POINT TRANSLATOR BETWEEN ARRAYS
0001          SUBROUTINE LOCATE(M, N, IM, IN, JM, JN)
      C  GIVEN THE ARRAY POINT (M,N) - LOCATE - FINDS THE CELL
      C  LOCATIONS (IM,JM) AND (IN,JN) ON THE ELECTRODE GRID
      C  AT PRESENT, THE ELECTRODE GRID IS (12X7) = 84 CELLS
0002          IM = (M-1)/12 + 1
0003          IN = (N-1)/12 + 1
0004          JM = M - 12*(IM - 1)
0005          JN = N - 12*(IN - 1)
0006          RETURN
0007          END
```

                            (c)

```
      C  COEFFICIENT CALCULATOR BASED UPON NUMERICAL INTERGRATIONS AND
      C  CURVE FITTING - SEE NOTES
0001          FUNCTION COEF(IX,IY,H)
0002          IF (H)  500, 10, 200
0003    10    IF (IX*IY - 1)  20, 50, 300
0004    20    IF (IX + IY - 1)  30, 40, 300
0005    30    COEF = .1403
0006          RETURN
0007    40    COEF = .0413
0008          RETURN
0009    50    COEF = .0288
0010          RETURN
0011    200   IF (IX+IY)  500,230,300
0012    230   COEF = 1./(12.563*H + 7.128*EXP(-.719*H))
0013          RETURN
0014    300   COEF = .0398/SQRT(IX**2 + IY**2 + H**2/4. )
0015    500   RETURN
0016          END
```

                            (d)

Figure 98 (*Continued*)

194

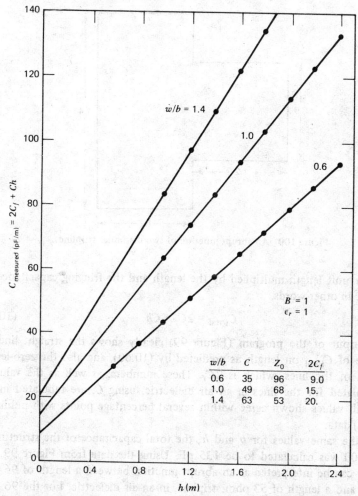

Figure 99 Capacitance versus length for various striplines, capacitance matrix technique.

to follow the formulation of the problem as described thus far, not for numerical or computer storage efficiency. Possible improvements in the program are discussed at the end of this section.

Figure 99 shows (graphically) the output of the program for several studies of simple rectangular strips, using a ground plane separation of $2h = B = 1$ and a cell size of $2b = 0.2$. Rectangular strips were studied of widths of 3, 5, and 7 cells, and lengths varying from 7 to 12 cells.

For a length of center conductor of stripline between two infinite planes, the total capacitance to ground should be the sum of the transmission line capaci-

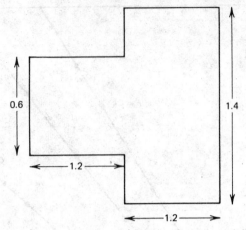

Figure 100 An abrupt junction of two dissimilar striplines.

tance per unit length multiplied by the length and the fringing capacitances at each end. In other words,

$$C_{\text{total}} = 2C_f + Ch \tag{10.60}$$

The output of the program (Figure 99) clearly shows the straight line dependence of C_{total} on length as predicted by (10.60), and also the zero-length intercept of the lines, which is $2C_f$. These numbers, as well as the value of Z_0 calculated for the lines in an air dielectric, using C, are tabulated in the figure. All values shown agree within several percentage points with published (analytic) data.

Using the same values for b and h, the total capacitance of the structure in Figure 100 was calculated to be 135 pF. Using the data from Figure 99, this structure can be interpreted as an abrupt junction between a length of 96 ohm stripline and a length of 53 ohm stripline, in an air dielectric. For the 96 ohm line, the capacitance of the exposed line edge plus the length of the line is $C_{96} = 9./2 + 1.2(35) = 46.5$. For the 53 ohm length, $C_{53} = 20/2 + 1.2(63) = 85.6$. The excess capacitance of the junction is therefore $135 - (46.5 + 85.6) \approx 3$ pF. A proper electrical circuit model for this abrupt junction would therefore consist of the two lines meeting at a point that is shunted by a 3 pF capacitor. If, in practice, this 3 pF capacitance is unacceptable for a given application, the computer program could be used to model a "trim" of the wider line near the junction, to decrease and possibly eliminate the excess capacitance of the junction.

In the computer program as shown, 7500 terms in the summation of (10.58) are taken to achieve reliable accuracy for arbitrary values of h/b. If this summation were carried out for each of the $84 \times 84 = 7056$ elements of the $L_{i,j}$ array, there would be an unnecessary waste of computer time. This is because although

$L_{i,j}$ contains 7056 numbers, only 63 of these numbers are different. There are 84 self-coupling ($L_{i,i}$) terms, some number of nearest neighbor ($L_{i,i+1}$) = ($L_{i,i-1}$) = ($L_{i+1,i}$) = $\cdots$ terms, and so on. For this reason, a "lookup table" is generated. The table stores each new value of $L_{i,j}$ along with the necessary geometric information as to its meaning. No new values of $L_{i,j}$ are calculated until the table has been searched to see whether the value under consideration has already been calculated. The program is also set up so that as long as h and b are never changed, the table is saved and reused as necessary. The data points of Figure 99, for example, are most efficiently calculated by taking the largest rectangle first. Since all the subsequent rectangles are subsets of the largest, the full table is generated during the first pass, and merely referred to from that point onward.

There are two important computational improvements that were not used in the program as described—the decision being made to have the matrices and vectors in the program correspond exactly to those in the analysis and to sacrifice computational efficiency for the sake of the example. These improvements are as follows:

1. The simple rectangles studied have fourfold symmetry. That is, only one-quarter of them need be studied. This means that the same array size could have been used with an increase in resolution, or the same resolution could have been had with smaller arrays.

2. Since all values of $L_{i,j}$ are available from the lookup table based on simple integer calculations on the values of i and j, there is really no need to store the array $L_{i,j}$. The simultaneous equation solving subroutine could have calculated the values IDELT(i, j) and JDELT(i, j) and then found the needed values of $L_{i,j}$ repetitively while iterating. This would no doubt cause some increase in computing time, but the program would be greatly freed from the available core storage of the computer. Note that since the Gauss-Seidel iterations converge very quickly even without overrelaxation, it is not important to avoid a reasonable increase in the complexity of the calculations of the terms of the iterations.

10.4 SUGGESTED FURTHER READING

1. W. Panofsky and M. Philips, *Classical Electricity and Magnetism*, Addison-Wesley, Reading, Mass. Derives Thomson's theorem, which is basic to the formulation of Section 10.1. Also discusses the fundamentals of Green's function solutions.

2. E. Yamashita and K. Atsuki, "Stripline with Rectangular Outer Conductor and 3 Dielectric Layers," *IEEE Transactions on Microwave Theory and Technique*, Vol. MTT-18, No. 5, May 1970. Extends the solution for stripline (Section 10.1) to be usable for boxed microstrip with multiple dielec-

trics. Also gives the algebra for a cubic assumed form for the charge distribution.

3. R. Harrington, "Matrix Methods for Field Problems," *Proceedings of the IEEE*, Vol. 55, No. 2, February 1967. The capacitance matrix approach is one form of a generalized theory of "the method of moments." This paper discusses the mathematical formulation of the general theory and some of its many uses.

4. A. Farrar and A. Taylor, "Matrix Methods for Microstrip Three-Dimensional Problems," *IEEE Transactions on Microwave Theory and Technique*, Vol. MTT-20, No. 8, August 1972. The procedure described here is essentially that of Section 10.3. However the authors derive the analytic solution for the integral required for $L_{i,j}$ and also the proper image charge series for the inhomogeneous dielectric (microstrip) problem. The concept of multitransmission line junctions is discussed, and numerous curves of results are presented.

11

Transmission Lines As Impedance Matching Networks

In general, when a length of transmission line is terminated in an impedance other than its own characteristic impedance, the input impedance of the line will be neither the terminating impedance nor the line's characteristic impedance. This property can be used to advantage when it is necessary to transform impedances. A lossless (or low loss) line can be used as a lossless (or low loss) matching network. The simplest form of transmission line matching network is the quarter-wave line. A quarter-wave line can transform a pure resistance into a different pure resistance, therefore fulfills many practical needs.

Unfortunately a line can be a quarter-wave long at only one frequency. The quarter-wave line transformer, however, can be broadbanded by cascading several different impedance sections together. The concept of cascading different impedance line sections together can be extended to that of a tapered line of nonuniform cross section. The tapered line has a continuously varying characteristic impedance and in many applications can outperform cascaded uniform sections.

Tightly coupled transmission lines can be fabricated by twisting together the center conductors of two wires and winding this twisted pair about a toroidal ferrite core. The resulting networks can be fabricated with the input and output grounds actually referring to the same potential—a very unusual property for transmission line networks. Torodial coupled line systems have transformerlike properties, as well as many practical and desirable characteristics. Some configurations may produce a balanced output from an unbalanced input, or vice versa. These structures, called transmission line *balun transformers*, also fulfill a very practical requirement.

11.1 IMPEDANCE MATCHING WITH
QUARTER-WAVE SECTIONS OF LINE

For a length of transmission line that is one quarter-wave long at some frequency,

$$Z_{in} = Z_0 \frac{Z_L \cos(\beta h) + jZ_0 \sin(\beta h)}{Z_0 \cos(\beta h) + jZ_L \sin(\beta h)}\bigg|_{\beta h = \pi/2} = \frac{Z_0^2}{Z_L} \qquad (11.1)$$

In other words, the quarter-wave line will match a load Z_L to a source Z_0^2/Z_L. This relationship holds exactly only at the frequencies at which the line is an odd number of quarter-waves long. Let the lowest of these frequencies be ω_0, where

$$\omega_0 = \frac{\pi v}{2h} \qquad (11.2)$$

For ω near ω_0, let

$$\Delta\omega \equiv \omega - \omega_0$$

$$\cos\omega \simeq -\frac{\Delta\omega h}{v}$$

$$\sin\omega \simeq 0$$

Then

$$Z_{in} \simeq Z_0 \frac{-xZ_L + jZ_0}{-xZ_0 + jZ_L} \qquad (11.3)$$

where $x \equiv \Delta\omega h/v$.

Normalizing both sides of (11.3) to Z_0, we write

$$Z_{in} = \frac{Z_L x - j}{x - jZ_L} \qquad (11.4)$$

If this is to match a (normalized) source $Z_s = 1/Z_L$,

$$\Gamma \equiv \frac{Z_{in} - Z_s}{Z_{in} + Z_s} = \frac{(Z_L^2 - 1)x}{(Z_L^2 + 1)x - j2Z_L} \qquad (11.5)$$

Evaluating the quality of the match in terms of the VSWR,

$$VSWR = \frac{1 + |\Gamma|}{1 - |\Gamma|} \qquad (11.6)$$

Figure 101 shows VSWR versus $\Delta\omega h/v$ for the cases $Z_L = 2$ and $Z_L = 4$.

The usefulness of an impedance matching transformer is probably best measured in terms of the transducer loss response of the transformer when driven by

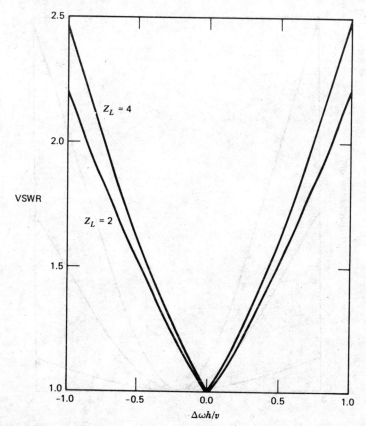

Figure 101   VSWR versus $\Delta\omega h/v$ for a quarter-wave matching section, $1:2$ and $1:4$ impedance matches.

and terminated in the desired (unequal) impedances. The transducer loss of a two-port network is defined as

$$TL = 10 \, \text{Log}_{10} \left[\frac{R_L^*}{4R_s} \left| \frac{V_s}{V_L} \right|^2 \right] \tag{11.7}$$

Transducer loss is a measure of how well the transformer (in this case) is doing its job as compared with an ideal transformer. This is a much more meaningful calculation than insertion loss in studies of impedance matching networks because in the case of impedance matching, even a poor transformer is often better than none at all. For that matter, a matching transformer can actually show an insertion gain!

Figure 102 plots the transducer loss versus frequency for three different quarter-wave matching sections. Each has a nominal center frequency of 1 and

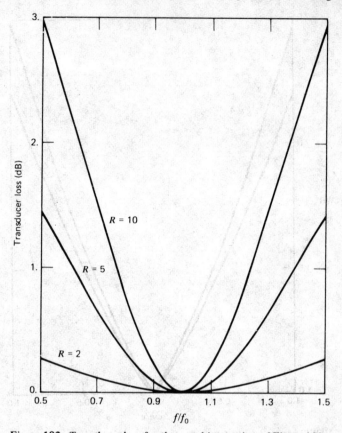

Figure 102 Transducer loss for the matching section of Figure 101.

each matches from 1 ohm to R ohms at center frequency. For R close to 1 there is no problem with bandwidth, but as R increases, the bandwidth of the transformer narrows drastically.

The useful bandwidth of transmission line matching sections can be improved by transforming impedance gradually, through two or more intermediate steps. In other words, two or more sections of line, each 1 quarter-wave long at some f_0 are cascaded. The choice of the intermediate impedance(s) will determine both the bandwidth and the shape of the frequency response curve. Figure 103 shows three examples of two-section transformers. Each is designed to match 1 ohm to 10 ohms, with a center frequency at $f_0 = 1$.

Multisection quarter-wave transformer synthesis has been examined in great detail in the literature, and some of the published results are referenced at the end of this chapter. These structures can be thought of, in general terms, as a type of bandpass filter. The transformer design parameters therefore closely

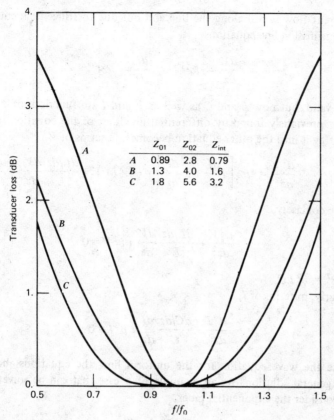

	Z_{01}	Z_{02}	Z_{int}
A	0.89	2.8	0.79
B	1.3	4.0	1.6
C	1.8	5.6	3.2

Figure 103 Transducer loss for several two-section, quarter-wave matching sections, 1:10 impedance match.

resemble filter design parameters such as 3 dB bandwidth, "passband" response, and skirt response.

11.2 TAPERED TRANSMISSION LINES

When the cross section of a transmission line is not constant, the definitions and relations developed thus far must be reexamined cautiously. Assuming a TEM system, the propagation velocity is not a function of the line impedance, therefore should not vary with position along the line. The characteristic impedance, on the other hand, is no longer a constant—and may not even have meaning in any general sense.

If the cross-sectional parameters of a line vary slowly enough with position that current flow is still along the line and not due to fringe field coupling, the basic transmission line equations

$$\frac{dV}{dz} = -j\omega LI, \qquad \frac{dI}{dz} = -j\omega CV$$

are still valid, but now L and C as well as V and I are functions of z. The wave equation previously found by differentiating either of the equations above and substituting it into the other equation becomes, in terms of V,

$$\frac{d^2 V}{dz^2} = -j\omega \left[I\frac{dL}{dz} + L\frac{dI}{dz} \right] = -\omega^2 LCV + \frac{dL/dz}{L}\frac{dV}{dz} \qquad (11.8)$$

or, more concisely,

$$\frac{d^2 V}{dz^2} - \frac{dL/dz}{L}\frac{dV}{dz} + \beta^2 V = 0$$

where $\beta^2 = \omega^2 LC$.

Similarly, in terms of I,

$$\frac{d^2 I}{dz^2} - \frac{dC/dz}{C}\frac{dI}{dz} + \beta^2 I = 0 \qquad (11.9)$$

Unlike the wave equation for the uniform line, the equations above do not have a general solution. As an example of a case that can be solved in closed form, consider the exponential taper,

$$L(z) = L_0 e^{qz} \qquad (11.9)$$

Since LC must be a constant in a true TEM system,

$$C(z) = C(z)e^{-qz} \qquad (11.10)$$

Although it is not clear what and even whether the term "characteristic impedance" has any real meaning in this situation, it is still useful to define $Z_0(z)$ as

$$Z_0(z) \equiv \sqrt{\frac{L}{C}} = \sqrt{\frac{L_0}{C_0}}\, e^{qz} = Z_0(0)e^{qz} \qquad (11.11)$$

The wave equations for V and I are, by direct substitution,

$$\frac{d^2 V}{dz^2} - q\frac{dV}{dz} + \beta^2 V = 0 \qquad (11.12a)$$

$$\frac{d^2 I}{dz^2} + q\frac{dI}{dz} + \beta^2 I = 0 \qquad (11.12b)$$

These equations can be solved in closed form, yielding

$$V(z) = e^{qz/2} \left[C_1 \exp\left(-\frac{\sqrt{q^2 - 4\beta^2}\ z}{2} \right) + C_2 \exp\left(\frac{\sqrt{q^2 - 4\beta^2}\ z}{2} \right) \right]$$

(11.13a)

$$I(z) = \frac{je^{-qz/2}}{\omega L_0} \left\{ \left[\frac{q}{2} - \frac{1}{2}\sqrt{q^2 - 4\beta^2} \right] C_1 \exp\left(-\frac{\sqrt{q^2 - 4\beta^2}\ z}{2} \right) \right.$$
$$\left. + \left[\frac{q}{2} + \frac{1}{2}\sqrt{q^2 - 4\beta^2} \right] C_2 \exp\left(\frac{\sqrt{q^2 - 4\beta^2}\ z}{2} \right) \right\}$$

(11.13b)

To put (11.13) into a more convenient and easily interpretable form, define β' as

$$\beta' = \tfrac{1}{2}\sqrt{4\beta^2 - q^2} = \tfrac{1}{2}\sqrt{4\omega^2 LC - q^2}$$

(11.14)

Then (11.13a) and (11.13b) can be rewritten as

$$V(z) = e^{qz/2}[C_1 e^{-j\beta'z} + C_2 e^{+j\beta'z}]$$

(11.15a)

$$I(z) = \frac{e^{-qz/2}}{\omega L_0} \cdot \left[C_1\left(\frac{jq}{2} + \beta' \right) e^{-j\beta'z} + C_2\left(\frac{jq}{2} - \beta' \right) e^{+j\beta'z} \right]$$

(11.15b)

As these equations indicate, $V(z)$ is the sum of two waves propagating in the $+z$ and $-z$ directions, as was the case for a uniform line, so long as β' is real. This means that there is a minimum frequency at which waves will propagate along this tapered line for any given absolute value of q. Equivalently, for a given frequency signal there is a maximum rate of taper, that is, a maximum absolute value of q, which will allow propagating signals.

For a propagating $V(z)$, V is exponentially either growing or falling with z, depending on the sign of q. This is reminiscent of wave propagation along a lossy line, but in this case the line was assumed to be lossless. In other words, either the line is gradually transforming the impedance—the ratio of the voltage to the current—or it is continually reflecting some signal, or both, with changing z.

If the line is terminated at $z = h$ with $Z_L = Z_0(h)$, then, from (11.15),

$$\sqrt{\frac{L_0}{C_0}}\, e^{qh} = \frac{\omega L_0 e^{qh/2}[C_1 e^{-j\beta'h} + C_2 e^{+j\beta'h}]}{e^{-qh/2}[C_1(jq/2 + \beta')e^{-j\beta'h} + C_2(jq/2 - \beta')e^{+j\beta'h}]}$$

(11.16)

This leads immediately to

$$C_2 = C_1 e^{-j2\beta'h} \left(\frac{jq/2 + \beta' - \beta}{-jq/2 + \beta' + \beta} \right)$$

(11.17)

and

$$Z_{in} = Z_0 \left[\frac{\beta' \cos(\beta'h) + j(\beta - jq/2)\sin(\beta'h)}{\beta' \cos(\beta'h) + j(\beta + jq/2)\sin(\beta'h)} \right] \tag{11.18}$$

The reflection coefficient as seen at the input ($z = 0$) is

$$\Gamma_{in} = \frac{q/2}{\beta' \cot(\beta'h) + j\beta} \tag{11.19}$$

Figure 104 plots the magnitude of Γ as a function of βh for $qh = 2$. Here for $\beta h = 0$, $|\Gamma| = 0.76$. This is perfectly reasonable, since this transformer should match

$$Z_0(h) = Z_0(0)e^2 = 7.39 Z_0(0)$$

and if the transformer is 0 wavelengths long (i.e., does not exist), $|\Gamma| = (7.39 - 1)/(7.39 + 1) = 0.76$. The figure shows an infinite number of ideal-match points, separated by regions of nonideal impedance match. As βh gets larger, the peak values of the mismatches get smaller. In other words, for a given minimum frequency of operation, and a given maximum allowable mismatch, the shortest possible exponentially tapered matching section can be specified. Obviously, the three criteria above can be permuted—that is, two of the three conditions (minimum operating frequency, maximum mismatch, minimum length) can be chosen independently and the third condition then calculated.

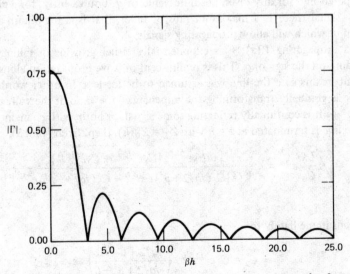

Figure 104 $|\Gamma|$ versus βh for an exponentially tapered line, $qh = 2$.

From (11.19) it can be seen that the peaks of mismatch occur at $\cot(\beta'h) = 0$. This means that they occur at $\beta'h = N\pi/2$, for N odd. At these peaks

$$|\Gamma|_{max} = \frac{q}{2\beta} = \frac{qh}{2\beta h} \quad \text{or} \quad qh = |\Gamma|_{max}(2\beta h) \tag{11.20}$$

Since

$$\frac{Z_0(h)}{Z_0(0)} = e^{qh} = \exp(2\beta h\,|\Gamma|_{max}) \tag{11.21}$$

$$(\beta h)\,|\Gamma|_{max} = \tfrac{1}{2}\,Ln\left|\frac{Z_0(h)}{Z_0(0)}\right| = \text{constant} \tag{11.22}$$

The relationship between βh and $|\Gamma|_{max}$ is therefore hyperbolic.

Using (11.11), (11.15), and (11.17), the transducer loss for the exponentially

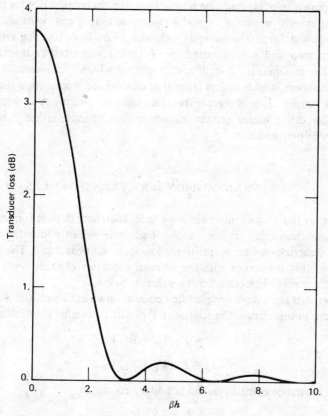

Figure 105 Transducer loss for the tapered line of Figure 104.

tapered line can be calculated. From the defining equation for transducer loss, (11.7),

$$\text{TL} = 10 \, \text{Log} \left[\frac{e^{qh}}{4} \left| \frac{V(0) + I(0)R_s}{V(h)} \right|^2 \right] \qquad (11.23)$$

where

$$V(0) = C_1 + C_2$$

$$V(h) = e^{qh/2} \left[C_1 e^{-j\beta'h} + C_2 e^{+j\beta'h} \right]$$

$$I(0)R_s = \frac{1}{\beta} \left[C_1 \left(\beta' + \frac{jq}{2} \right) + C_2 \left(-\beta + \frac{jq}{2} \right) \right]$$

Figure 105 shows transducer loss versus βh for $qh = 2$. As would be expected from the equation for Γ, the transducer loss repeatedly touches zero and the peak value of the loss between adjacent zeros decreases with increasing βh.

The exponentially tapered line is not the optimum tapered line for most applications. It merely serves as a good example in that it can be handled conveniently in closed form. Optimum tapers, using various optimizing criteria, have been calculated and are discussed in the references cited at the end of this chapter. The exponential taper does exemplify the basic characteristic of tapered line transformers, which differs from that of single or multisection quarter-wave line transformers. This characteristic is a minimum frequency at which a given transformer can produce certain minimum performance, rather than a center frequency of performance.

11.3 TRANSMISSION LINE TRANSFORMERS

Transmission line transformers are impedance matching devices formed by winding coupled transmission lines about high permeability toroidal cores. The resulting structures are small, relatively low loss, and broadband. The interaction of coupled line properties with the unusual geometry of these units allows for these performance characteristics, as is shown below.

Consider first the simple coupled line circuit shown in Figure 106. Since V_2 and V_4 are zero in this circuit, the four-port Y matrix (assuming identical lines) is

$$I_1 = Y_{11} V_1 + Y_{13} V_3 \qquad (11.24a)$$

$$I_3 = Y_{13} V_1 + Y_{33} V_3 \qquad (11.24b)$$

The load resistance contributes the boundary condition

$$I_3 = -Y_L V_3 \qquad (11.25)$$

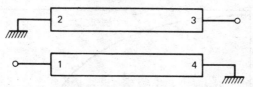

Figure 106 Coupled line circuit for toroidal transformer model.

and from these equations we can derive the voltage gain of the coupled line circuit,

$$A_V = \frac{-Y_{13}}{Y_{11} + Y_L} \tag{11.26}$$

Referring to (8.40), the relation above becomes

$$A_V = \frac{-j(Y_{0o} - Y_{0e}) \csc(\beta h)}{-j(Y_{0o} + Y_{0e}) \cot(\beta h) + 2Y_L} \tag{11.27}$$

Equation 11.27 can be separated into a magnitude and an angle,

$$|A_V|^2 = \frac{(Y_{0o} - Y_{0e})^2 \csc^2(\beta h)}{(Y_{0o} + Y_{0e})^2 \cot^2(\beta h) + 4Y_L^2} \tag{11.28}$$

$$\arg(A_V) = \frac{\pi}{2} + \tan^{-1}\left[\frac{(Y_{0o} + Y_{0e}) \cot(\beta h)}{2Y_L}\right] \tag{11.29}$$

Assuming that the lines are very tightly coupled, in the terminology of Chapter 8, $R \equiv Y_{0o}/Y_{0e} \gg 1$. Since this circuit is identical to that of the bandpass filter example in Chapter 8, we would expect A_V to be a bandpass response, and for that matter, a very broadband bandpass response. Rewriting the two equations above in terms of this approximation,

$$|A_V|^2 = \frac{1}{\cos^2(\beta h) + (2Y_L \sin(\beta h)/Y_{0o})^2} \tag{11.30}$$

$$\arg(A_V) = \frac{\pi}{2} + \tan^{-1}\left(\frac{Y_{0o} \cot(\beta h)}{2Y_L}\right) \tag{11.31}$$

If

$$Y_L = \frac{Y_{0o}}{2} \tag{11.32a}$$

or, equivalently,

$$Z_L = \frac{2}{Z_{0o}} \tag{11.32b}$$

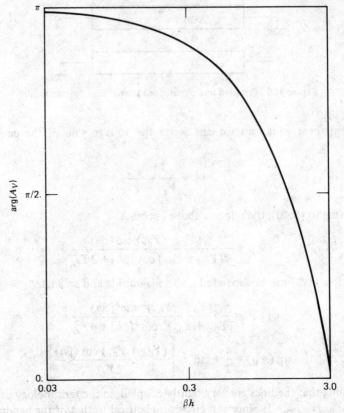

Figure 107 Phase response of the toroidal inverting transformer.

the voltage gain magnitude and angle approximations become

$$|A_V|^2 = 1 \qquad (11.33)$$

$$\arg (A_V) = \frac{\pi}{2} + \tan^{-1} \, [\cot (\beta h)] = \pi - \beta h \qquad (11.34)$$

Figure 107 plots the phase response (11.34) versus frequency (on a semi-log scale).

The same circuit that was analyzed previously as a bandpass filter has become, under certain approximations, a unity gain–phase inverting transformer. In principle it is useful from dc until βh approaches $\pi/2$. In practice the bandwidth is usually smaller than this, but still is quite large in a well-designed and well-constructed unit.

The form the inverting transformer usually takes in practice is that of a twisted pair of wires looping several times through a high permeability toroid, as in

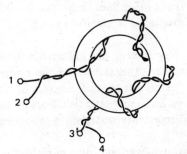

Figure 108 Construction of the toroidal inverting transformer.

Figure 108. However Figure 108 does not show the ground return for the coupled pair of lines. In practice, of course, the ground return must exist.

To measure Y_{oo} for the structure of Figure 108, set $V_1 = -V_2 = V_i$, and terminate ports 3 and 4 to ground in nonreflecting loads (Figure 109). Actually, the value of load impedances required to realize this termination has not been specified. Assume for the moment that they have been found—by trial and error if necessary.

Since the two wires are very tightly coupled (twisted together), the input admittance between either port and ground is determined almost entirely by the wires themselves, and hardly at all by the ground return. Fortunately, in practice, the ground return is usually very far away compared with the distance between the twisted wires. In other words, the electric field lines travel almost exclusively from line to line rather than from line to ground. If Y_0 is defined as the characteristic admittance of the twisted pair of wires when treated as an ordinary two-wire transmission line, then because Y_{oo} is measured from one line to ground, we write

$$Y_{oo} \simeq 2Y_0 \qquad (11.35)$$

To measure Y_{oe}, set $V_1 = V_2 = V_i$ and again measure the admittance to ground from either port 1 or port 2 (Figure 109). In this case none of the elec-

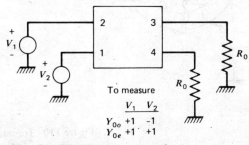

Figure 109 Y_{oo} and Y_{oe} measurement circuits for a pair of lines.

tric field lines from one line may terminate at the other line. They must "seek" the far ground. The symmetric mode line capacitance is therefore much lower than the asymmetric mode capacitance.

The high permeability toroid core causes the symmetric line inductance to be much higher than the asymmetric line inductance. This is because the return currents in the symmetric case must flow in the distant ground, whereas in the asymmetric case they flow in the "other" wire. Since magnetic field lines in the symmetric case must pass through the high permeability material, the resulting inductance is comparatively high.

The relationship resulting from the comparative even and odd mode inductances and capacitances just described is that

$$Y_{0e} = \sqrt{\frac{C_{\text{even}}}{L_{\text{even}}}} \ll Y_{0o} = \sqrt{\frac{C_{\text{odd}}}{L_{\text{odd}}}} \tag{11.36}$$

In summary, the twisted pair of wires looping through the toroid approximates the conditions:

1. $Y_{0o} = 2Y_0$ due to the twisted pair of wires.
2. $Y_{0e} \ll Y_{0o}$ due to the geometry and the toroid.

These two conditions are identical to (11.35) and the condition $R \gg 1$. Therefore this structure should behave like an inverting transformer, with response as shown in Figure 107.

The same structure described for use as an inverting transformer can be used entirely differently by connecting the leads as in Figure 110. In this case the

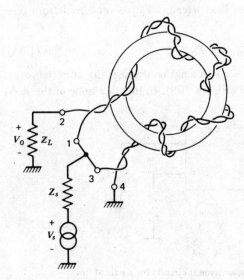

Figure 110 Toroidal transformer connections for a 1:4 impedance match.

terminal conditions for the four-port matrix are

$$V_4 = 0$$
$$V_1 = V_3 = V_i \qquad\qquad (11.37)$$
$$V_2 = V_0$$

The nodal circuit equations are then

$$I_1 = (Y_{11} + Y_{13})V_i + Y_{12}V_0$$
$$I_2 = (Y_{12} + Y_{14})V_i + Y_{11}V_0 \qquad\qquad (11.38)$$
$$I_3 = (Y_{11} + Y_{13})V_i + Y_{14}V_0$$

Also, by inspection,

$$I_i = I_1 + I_2$$
$$I_0 = I_2 \qquad\qquad (11.39)$$

so that (11.38) becomes

$$I_i = 2(Y_{11} + Y_{13})V_i + (Y_{12} + Y_{14})V_0$$
$$I_0 = (Y_{12} + Y_{14})V_i + Y_{11}V_0 \qquad\qquad (11.40)$$

Let $\bar{y}_{11}, \bar{y}_{12}, \bar{y}_{22}$, be the two-port Y parameters for the circuit. The equations above become

$$I_i = \bar{y}_{11} V_i + \bar{y}_{12} V_0$$
$$I_0 = \bar{y}_{12} V_i + \bar{y}_{22} V_0 \qquad\qquad (11.41)$$

where

$$\bar{y}_{11} = 2(Y_{11} + Y_{13}) \simeq -j2Y_0 \left[\cot (\beta h) + \csc (\beta h)\right] \qquad (11.42a)$$

$$\bar{y}_{12} = Y_{12} + Y_{14} \simeq -j\tfrac{1}{2} \left[(Y_{0o} - Y_{0e}) \cot (\beta h) + (Y_{0o} + Y_{0e}) \csc (\beta h)\right]$$
$$\qquad\qquad (11.42b)$$

$$\bar{y}_{22} = Y_{11} \simeq -jY_0 \cot (\beta h) \qquad\qquad (11.42c)$$

Using these, the voltage gain of the network is

$$A_V = \frac{-Y_0 \left[\cos (\beta h) + 1\right]}{Y_0 \cos (\beta h) + jY_L \sin (\beta h)} \qquad\qquad (11.43)$$

If $Y_0 = Y_L$, then

$$|A_V| = \cos (\beta h) + 1 \qquad\qquad (11.44)$$

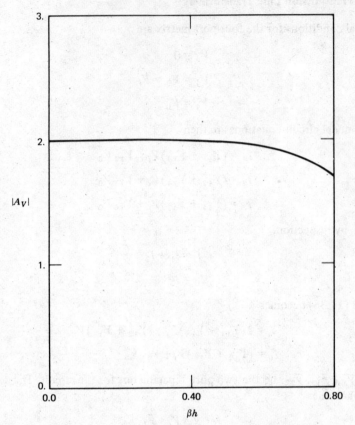

Figure 111 A_V versus βh for the transformer of Figure 110.

Figure 111 plots (11.44). Note that at low frequencies, $|A_V| = 2$. This usually implies a $1:4$ impedance match. Calculating Z_{in} yields

$$|Z_{in}| = \left| \frac{\bar{y}_{22} + Y_L}{\det(\bar{y}) \, 4\bar{y}_{11} Y_L} \right| = \frac{1}{Y_0} \frac{1}{[4(\cos(\beta h) + 1)^2 + \sin^2(\beta h)]^{1/2}}$$

$$(11.45)$$

This structure is indeed a $1:4$ impedance matching transformer—as Figure 112 shows.

There are many possible variations in the design of transmission line transformers of this type. Only two have been shown here. Some of the combinations involve three, rather than two, wires. One of these is described in the next section. The references at the end of this chapter contain more detailed analysis

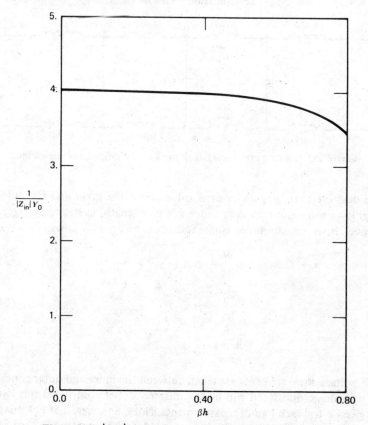

Figure 112 $|Z_{in}|$ versus βh for the transformer of Figure 110.

of the different types of toroidal transmission line transformers as well as some practical construction technique descriptions.

11.4   UNBALANCED CIRCUITS, GROUNDING, AND BALUN TRANSFORMERS

The simple circuit model for an unbalanced transmission line, from which the transmission line equations were derived in Chapter 1, is in a sense misleading. It is very convenient to assume that all the inductance of a line is in the "ungrounded" conductor, and this assumption is mathematically accurate. It may, unfortunately, lead to misimpressions about the actual voltage drop along the "grounded" conductor.

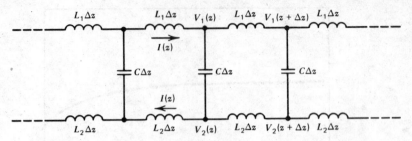

Figure 113 Exact incremental circuit model for a single transmission line.

Consider, for example, the coaxial cable. Both the inner and the outer conductors have some nonzero inductance per unit length, as shown in Figure 113. The three circuit equations describing this situation are as follows:

$$\frac{\partial i}{\partial z} = -j\omega C(V_1 - V_2) \qquad (11.46a)$$

$$\frac{\partial V_1}{\partial z} = -j\omega L_1 i \qquad (11.46b)$$

$$\frac{\partial V_2}{\partial z} = -j\omega L_2 i \qquad (11.46c)$$

If all voltage measurements are taken between the inner and outer conductors, the only voltages measured will be the differences between $V_1(z)$ and $V_2(z)$—at some given z for each pair of measurements. Note, however, that (11.46a) refers only to the difference $V_1 - V_2$. Subtracting (11.46c) from (11.46b), we have

$$\frac{\partial(V_1 - V_2)}{\partial z} = -j\omega(L_1 + L_2)i \qquad (11.47)$$

Equations 11.47 and 11.46a form a pair of equations that completely describe the system in terms of $V_1 - V_2$. Note that in these equations the inductance terms appear only as the sum $L_1 + L_2$. It is therefore mathematically correct to model the transmission line as having the total inductance per unit length appearing in the center conductor and the lower nodes referenced to zero volts. This does not mean, however, that in a physical line there is no inductance in the outer conductor or that in a physical line there is no voltage drop along the outer conductor. In many practical cases $L_2 \ll L_1$ and for a short line length at low frequencies the foregoing statements can be taken as reasonable approximations.

At this point we appear to have lost the very useful concept of an unbalanced line with one conductor always at some "ground" potential. This concept is still useful as long as it is noted that in a transmission line network a ground refercnce is valid only at the particular value of z for which it is defined. Since there is no zero inductance path between two points, it is never correct to assume that two separated points are at the same potential—even though a transmission line circuit model that yields the correct circuit equations seems to imply that the two points are at the same potential. Since most networks never require the comparison of two different values of z (cross-sectional reference planes), the discrepancy rarely arises on paper. In practice it is often noted that "grounding" does not always have quite the desired results. Microwave circuit engineers refer to the situation just described by postulating "box currents"—that is, the currents that flow around the outside of supposedly shielded networks in the "shield" and cause various spurious voltage drops.

One of the very useful properties of the toroidally wound transmission line transformers described in the preceding section is that these transformers allow the definition of a unique ground reference at two different values of z. This was implied in the derivation of the properties of the inverting transformer without being mentioned explicitly. The toroid transformers accomplish this feat because the twisted pair lines, after being wrapped through the toroid several times, begin and end at the same point in space. The grounded wires from the input and the output are then tied together through an (essentially) zero length path, and their junction can be considered to be the ground reference for both the input and the output of the line.

A network that takes an unbalanced transmission line as an input and provides a balanced output is known as a *balun transformer*. Any balun transformer, of course, can be used in the reverse sense to the definition above. Balun transformers take many forms. From audio to low rf frequencies a balun transformer may simply be a lumped inductance $1:1$ transformer connected as in Figure 114. If the size of the transformer is very small compared with the wavelength of the highest frequency signal present, the two ground points in Figure 114 can be considered to be a single node.

At rf and microwave frequencies a balun transformer can be constructed using coupled transmission line sections. Many different circuits will perform the balun transformer function. One such circuit (Figure 115) utilizes dual con-

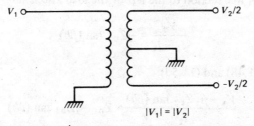

$$|V_1| = |V_2|$$

Figure 114 A low frequency balun transformer.

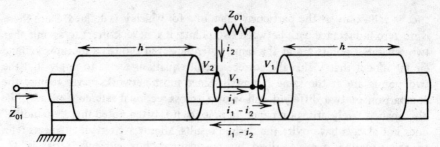

Figure 115 A transmission line balun transformer.

centric coaxial cables. In practice, the outer shield is often realized by the box in which the balun transformer is built.

The circuit of Figure 115 can be analyzed using the four-port coupled line matrix for the dual concentric cable derived in Chapter 8. In terms of understanding the mechanisms at work, however, it is convenient at this point to ignore the previous derivation and start over again in terms of the variables of this circuit. Assume that the region in the center of the balun, as shown, is electrically small enough that the interconnections between the transmission line sections and the load impedance can be considered to be an ideal lumped circuit. Referring to the voltages and currents as labeled in Figure 115, across the load resistance,

$$\frac{V_1 - V_2}{i_2} = Z_{01} \tag{11.48}$$

The transmission line section to the right of the load is a simple coaxial line, made identical to the inner and outer shields of the dual concentric line shown to the left of the load. If the characteristic impedance of this (right-hand) line is Z_{02}, since the line is shorted at its far end,

$$\frac{V_1}{i_1 - i_2} = -jZ_{02} \tan{(\beta h)} \tag{11.49}$$

Since the outer line section to the left of the load also is shorted at its far end,

$$\frac{V_2}{i_2 - i_1} = -jZ_{02} \tan{(\beta h)} \tag{11.50}$$

Combining (11.48) and (11.50),

$$\frac{V_1 - ji_1 Z_{02} \tan{(\beta h)}}{i_2} = Z_{01} - jZ_{02} \tan{(\beta h)} \tag{11.51}$$

Using (11.49),

$$\frac{V_1 - ji_1 Z_{02} \tan(\beta h)}{i_2} = Z_{01} - jZ_{02} \tan(\beta h) \tag{11.52a}$$

which reduces to

$$\frac{V_1}{i_1} = \frac{Z_{01}}{2 - Z_{01}/[jZ_{02} \tan(\beta h)]} \tag{11.52b}$$

Substituting the result above into (11.50) and (11.51),

$$\frac{V_2}{i_2} = -\frac{Z_{01}}{2} \tag{11.53}$$

$$\frac{i_2}{i_1} = \frac{j2Z_{02} \tan(\beta h)}{j2Z_{02} \tan(\beta h) - Z_{01}} \tag{11.54}$$

From (11.49) and (11.50) it is apparent that

$$V_2 = -V_1 \tag{11.55}$$

In other words, the voltage appearing across the load resistance Z_{01} is a balanced voltage—symmetric about a ground reference point at the center of the balun. This means that Z_{01} can be replaced by a symmetric balanced line of characteristic impedance Z_{01}, if desired.

The inner coaxial line to the left of the load has a characteristic impedance Z_{01}. At the center of the balun this line sees a load impedance given by

$$\begin{aligned}
Z_L &= \frac{V_1 - V_2}{i_1} = \frac{V_i}{i_1} - \frac{V_2}{i_1} = \frac{V_1}{i_1} - \frac{V_2}{i_2}\frac{i_2}{i_1} \\
&= Z_{01} \left[\frac{j2Z_{02} \tan(\beta h)}{j2Z_{02} \tan(\beta h) - Z_{01}} \right]
\end{aligned} \tag{11.56}$$

The input impedance at the unbalanced port of the balun transformer can be found by reflecting Z_L back through the length of Z_{01} line to the input port. Since such a calculation will affect only the magnitude of Z_{in}, the magnitude of V_{out}, the VSWR, and the insertion loss of the balun can be calculated directly using Z_L. Since the transformer itself is assumed to be lossless, any power going into it must be coming out into the balanced load. The insertion loss is therefore simply

$$IL = -10 \, \mathrm{Log}_{10} \left[\frac{4(Z_{02}/Z_{01})^2 \tan^2 \beta h}{4(Z_{02}/Z_{01})^2 \tan^2 \beta h + 1} \right] \tag{11.57}$$

Figure 116 plots (11.57) versus frequency for several real values of Z_{02}/Z_{01}. Clearly bandwidth is increased by keeping Z_{02}/Z_{01} large. In practice this means

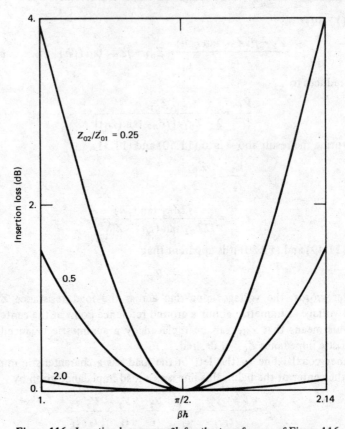

Figure 116 Insertion loss versus βh for the transformer of Figure 116.

that the outer conductor of the dual concentric cable must be as large as possible. This configuration is often achieved by constructing the balun transformer inside an oversized metal box and using the walls of the box themselves as the outer conductor.

The same balun circuit discussed previously can be realized in a very different structural form by noting that the right-hand line in Figure 116, shorted at its far end, can be "wrapped around" and brought back so that the shorted end is at the same physical point as the balun input ground (Figure 117a). If the box enclosing the balun is used as the outer conductor, and not explicitly shown in the circuit, the balun resembles Figure 117b. Finally, replacing the inner coaxial line with a twisted pair line, and wrapping both the twisted pair line and the "right-hand" line about a toroid to increase the "even" or "common" mode inductances, leads to the toroidal transformer form of the balun (Figure 117c).

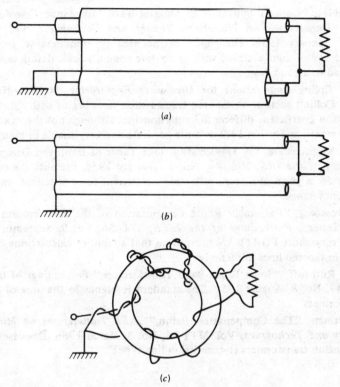

Figure 117 Evolution of the toroidal transformer balun transformer from the transmission line balun transformer.

As in the case of the toroidal transformers discussed in Section 11.3, there are many variations on the transmission line balun transformer circuit. The principle of operation of them all must, by nature, be identical to the structure described here. This tends to be elusive, however, because often the circuit disguises the ground return line well, many times simply not showing it at all. The clue to untangling these circuits is to examine every length of transmission line shown, and if at the end of any line the currents in the two wires are not exactly equal and opposite, there is a third current path not being shown.

11.5 SUGGESTED FURTHER READING

1. S. Cohn, "Optimum Design of Stepped Transmission-Line Transformers," *IEEE Transactions on Microwave Theory and Techniques*, Vol. MTT-3, No. 2, February 1955.

2. H. Riblet, "General Synthesis of Quarter-Wave Impedance Transformers," *IEEE Transactions on Microwave Theory and Techniques*, Vol. MTT-5, No. 1, January 1957. The papers in this area are innumerable. In general, almost any of them coupled with up-to-date computer calculating capabilities can lead to a good design.

3. R. E. Collin, *Foundations for Microwave Engineering*, McGraw-Hill, New York. Collin's section on tapered transmission lines begins with a distributed reflection coefficient differential equation that although not the exact analysis presented in Section 11.2, is quite accurate and very usable in many cases.

4. R. W. Klopfenstein, "A Transmission Line Taper of Improved Design," *Proceedings of the IRE*, Vol. 44, No. 1, January 1956. Presents the optimum design for a given minimum reflection coefficient for a specified length and frequency range.

5. M. Grossberg, "Extremely Rapid Computation of the Klopfenstein Impedance Taper," *Proceedings of the IEEE*, Vol. 56, No. 9, September 1968. Gives very short FORTRAN subroutine that simplifies calculations of Klopfenstein tapered lines immensely.

6. C. L. Ruthroff, "Some Broad-Band Transformers," *Proceedings of the IRE*, Vol. 47, No. 8, August 1959. The standard reference in the area of toroidal transformers.

7. G. Oltman, "The Compensated Balun," *IEEE Transactions on Microwave Theory and Techniques*, Vol. MTT-14, No. 3, March 1966. Discusses broadband balun transformers (transmission line type).

12

Potpourri: Several Uses of Transmission Lines as Circuit Elements

Since a lossless transmission line that is either open- or short-circuited at its far end appears as a reactance at its input, transmission line sections can be used in many applications as substitutes for lumped reactances. Both inductive and capacitive reactances may be realized. Also, since a section of line will go through both short- and open-circuit input impedance values as frequency is varied, transmission line sections can be used to substitute for both parallel and series resonant circuits.

In some applications—typically at VHF and at higher frequencies—it is very difficult to build conventional networks such as matched attenuators (pads) and amplifiers without taking into account the transmission line properties of the lines connecting the components in these networks. Fortunately it is usually possible to design these networks with the transmission line nature of the interconnections in mind and to take advantage of this nature. Various circuit elements such as bias T's, dc returns, and dc blocks emerge when this is done. Also, certain networks such as the transmission line–PIN diode absorption attenuator become very practical at UHF and at microwave frequencies because the necessary transmission line sections in these networks become short and easily attainable.

12.1 IMPEDANCE APPROXIMATIONS FOR LOW LOSS LINES

Let us define a low loss transmission line as one that, over some frequency range of interest, satisfies

$$R \ll j\omega L \tag{12.1}$$

223

$$G \ll j\omega C \tag{12.2}$$

Now, by definition,

$$\gamma = [(R + j\omega L)(G + j\omega C)]^{1/2}$$

$$= \omega\sqrt{LC}\left[-1 + \frac{RG}{\omega^2 LC} + j\left(\frac{R}{\omega L} + \frac{G}{\omega C}\right)\right]^{1/2} \tag{12.3}$$

The second term in brackets in (12.3) is of the second order in smallness, using (12.1) and (12.2). Dropping this term leaves

$$\gamma \simeq \omega\sqrt{LC}\left[-1 + j\left(\frac{R}{\omega L} + \frac{G}{\omega C}\right)\right]^{1/2} \tag{12.4}$$

In (12.4) the imaginary term is much smaller than the real term. Therefore it can be represented accurately by the first two terms of a binomial expansion:

$$\gamma \simeq \omega\sqrt{LC}\left[j + \tfrac{1}{2}\left(\frac{R}{\omega L} + \frac{G}{\omega C}\right)\right] \tag{12.5}$$

To a first-order approximation, therefore, the loss term is

$$\alpha = \tfrac{1}{2}\left[R\sqrt{\frac{C}{L}} + G\sqrt{\frac{L}{C}}\right] \tag{12.6}$$

The phase term, to a first-order approximation, is not different from the lossless case

$$\beta = \omega\sqrt{LC} \tag{12.7}$$

The characteristic impedance of the line is given by

$$Z_0 = \sqrt{\frac{R + j\omega L}{G + j\omega C}} = \sqrt{\frac{L}{C}}\left[\frac{1 + R/j\omega L}{1 + G/j\omega C}\right]^{1/2} \approx \sqrt{\frac{L}{C}}\left[\left(1 + \frac{R}{j\omega L}\right)\left(1 - \frac{G}{j\omega C}\right)\right]^{1/2} \tag{12.8}$$

where, again, terms higher than first order have been dropped.

Following the procedure used in approximation γ,

$$Z_0 \simeq \sqrt{\frac{L}{C}}\left[1 + \frac{j}{2}\left(\frac{G}{\omega C} - \frac{R}{\omega L}\right)\right] \tag{12.9}$$

Note that Z_0 has a first-order imaginary component that in general should not be ignored.

The input impedance of an open-circuited line is

$$Z_{in} = -Z_0 \coth(\gamma h) \tag{12.10}$$

If the line is short enough and/or the loss is low enough that $\alpha h \ll 1$, then

$$Z_{in} \simeq Z_0 \frac{\cos{(\beta h)} + j\alpha h \sin{(\beta h)}}{\alpha h \cos{(\beta h)} + j \sin{(\beta h)}} \qquad (12.11)$$

which further simplifies to

$$Z_{in} \simeq Z_0 \frac{\alpha h - j \cos{(\beta h)} \sin{(\beta h)}}{(\alpha h \cos{(\beta h)})^2 + \sin^2{(\beta h)}} \qquad (12.12)$$

For a very short line,

$$Z_{in} \simeq Z_0 \left[\frac{\alpha - j\beta}{h \, (\alpha^2 + \beta^2)} \right] \qquad (12.13)$$

Using the values of Z_0 and α derived above, and further simplifying the resulting expression, we have

$$Z_{in} \simeq \frac{G}{h\omega^2 C^2} - j \frac{1}{\omega h c} \qquad (12.14)$$

Inverting this expression yields

$$Y_{in} \simeq hG + jh\omega C \qquad (12.15)$$

This result is intuitively expected. For a very short open-circuited line, the input admittance is simply the sum of the input capacitance and dielectric loss terms. The series resistance (R) simply does not enter into the picture.

For $\beta h \approx \pi/2$, (12.12) becomes

$$Z_{in} \simeq Z_0 \left[\alpha h + j \left(\beta h - \frac{\pi}{2} \right) \right] \qquad (12.16)$$

Substituting the derived expressions for Z_0, α, and β, into (12.16), Z_{in} has a real part

$$\text{Re}\, \{Z_{in}\} = \frac{h}{2} \sqrt{\frac{L}{C}} \left[R \sqrt{\frac{C}{L}} + G \sqrt{\frac{L}{C}} - \sqrt{LC} - \frac{\pi}{2\omega h} \right] \qquad (12.17)$$

and an imaginary part

$$\text{Im}\, \{Z_{in}\} = \sqrt{\frac{L}{C}} \left[\frac{h}{4\omega} k + \omega \sqrt{LC}\, h - \frac{\pi}{2} \right] \qquad (12.18)$$

where

$$k = \frac{G^2}{C} \sqrt{\frac{L}{C}} - \frac{R^2}{L} \sqrt{\frac{C}{L}} \qquad (12.19)$$

The assumption of a low loss line leads to the assertion that k must be small. In this case (12.19) will have a root given by

$$\omega_0 \simeq \frac{\pi}{2\sqrt{LC}\,h}\left[1 - \frac{2\sqrt{LC}\,h^2 k}{\pi^2}\right] \tag{12.20}$$

The frequency of the root is different from that of a lossless line by the factor $1 - 2\sqrt{LC}\,h^2 k/\pi^2$. Note that the resonant frequency (root) may be higher than, lower than, or even equal to that of the lossless line, depending on the values of the terms that make up k.

The derivative of the imaginary part of Z_0 with respect to ω, evaluated at the resonant frequency, ω_0, is

$$\left.\frac{dx}{d\omega}\right|_{\omega_0} \simeq Lh\left[1 - \frac{hk}{2\pi}\right] \tag{12.21}$$

Since this derivative is positive, this line is behaving much like a simple series R-L-C circuit near resonance. Since the line is approximately one-quarter wavelength long at resonance, it is usually referred to as a "quarter-wave" open-circuit resonator.

A series R-L-C lumped circuit has a resonant frequency given by

$$\omega_0^2 = \frac{1}{L_{\text{lumped}}C_{\text{lumped}}} \tag{12.22}$$

and at this resonant frequency,

$$\left.\frac{dZ_{\text{in}}}{d\omega}\right|_{\omega_0} = j2L_{\text{lumped}} \tag{12.23}$$

The real part of the input impedance of this lumped circuit is simply R_{lumped}. Equations 12.17, 12.20, and 12.21, together with (12.22) and (12.23), comprise equivalency relations for approximating a series R-L-C network with a quarter-wave open-circuit resonator. The resonator Q in this case is given by

$$Q = \frac{\omega_0 L_{\text{lumped}}}{R_{\text{lumped}}} \tag{12.24}$$

For a length of line short-circuited at the far end,

$$Z_{\text{in}} = Z_0 \tanh(\gamma h) \tag{12.25}$$

and using the low loss approximations,

$$Z_{\text{in}} \simeq Z_0 \frac{\alpha h + j\sin(\beta h)\cos(\beta h)}{\cos^2(\beta h) + (\alpha h\sin(\beta h))^2} \tag{12.26}$$

For a short length of line,

$$Z_{in} \simeq (\alpha h + j\beta h) \tag{12.27}$$

$$\simeq Rh + jh\omega L \tag{12.28}$$

Again, the result is not surprising. A very short length of line, shorted at the far end, looks like a series R-L circuit with the resistance and inductance values being determined by the line's resistance and inductance per unit length and the length.

The quarter-wave shorted line will have a resonance closely related to that of a parallel G-L-C lumped network, with the actual frequency of resonance again given by (12.20).

12.2 DC BLOCKS, DC RETURNS, AND BIAS T'S

Before discussing transmission line networks for either isolating or shorting dc paths on transmission lines, it is useful to introduce a circuit element that needs such functions. For the sake of simplicity, a two-wire device, the PIN diode, has been chosen. The basic considerations extend directly to three-wire devices such as the various semiconductor transistors (BJT, FET, etc.), which are useful at UHF and at higher frequencies.

A PIN diode is a semiconductor device whose rf conductance, over a large dynamic range, is directly proportional to the dc (bias) current flowing through the diode. The PIN diode can be made very small, thereby showing very slight parasitic effects over a frequency range from several megahertz to several gigahertz. Consequently, the PIN diode is a very useful rf switch and/or variable attenuator device at frequencies where conventional lumped circuit switches and variable resistances are all but useless.

To take advantage of the PIN diode it is necessary to be able to direct dc currents through an rf circuit with as little interaction between the dc and rf currents as possible. At low enough frequencies (approximately below 1 GHz), conventional lumped capacitors, inductors, and resistors perform these circuit functions adequately. At UHF and at microwave frequencies, unfortunately, most lumped element structures have properties dominated by parasitic effects and are simply inadequate for the needed circuital functions.

As an example of a transmission line rf device using a PIN diode, consider a simple SPDT switch (Figure 118). The switch can be realized at microwave frequencies by using two PIN diodes and arranging to turn one diode "on" (full current flowing) while the other is "off" (no current flowing). This can be accomplished easily by connecting the diodes as Figure 119. In this circuit a single

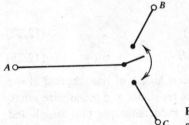

Figure 118 Schematic representation of an SPDT switch.

bias supply, switching point A from positive to negative, will turn one diode on while the other is off, and vice versa, provided there are return paths through points B and C. Unfortunately the rf circuit in which this SPDT switch is to be used usually cannot be relied on to provide these return paths.

One method of providing the dc return paths to points B and C is to connect, at each point, a quarter-wave transmission line that is grounded at its far end. Obviously dc current will flow freely through these quarter-wave lines, and points B and C will see a very high impedance "looking into" them.

If single frequency use only is contemplated for the switch, it is a straight-forward procedure to choose the length of the quarter-wave lines. If some measure of broadbanding is desired, design tradeoffs must be made. The input reactance to a shorted (lossless) quarter-wave line is $jZ_0 \tan(\beta h)$. For Z_0 as high as possible, the impedance of the dc return lines will stay high for some range of frequencies about the operating frequency. If acceptable limits to shunt reactance of the dc return lines are specified, then the characteristic impedance of the dc return lines can be calculated for a given frequency range about some nominal center frequency. The multistep quarter-wave transformers discussed in Chapter 11 can also be applied to this problem, but in many cases they are simply too lengthy to be practical.

Typically, a structure such as the SPDT switch is fabricated on a strip- or microstrip line, with the signal carrying lines and the dc return lines formed by

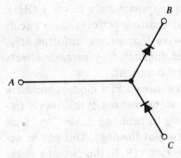

(Transmission line ground returns not shown)

Figure 119 SPDT PIN diode switch: rf signal path only.

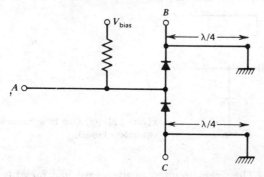

Figure 120 SPDT PIN diode switch showing the dc returns.

metal conductors on the same dielectric substrate. In this case the dielectric material should be chosen so that both the signal lines and the (usually thinner) dc return lines have realizable widths.

The problem of feeding dc bias to point A has not yet been discussed. There are two issues here. First, the dc path must present a high impedance to the rf line. Second, it is usually desirable to provide a dc "block" so that the bias voltage does not appear at the rf port of the switch.

The first requirement can be met by simply supplying the dc bias through a series resistor whose resistance is much larger than the characteristic impedance of the line (Figure 120). Although low frequency lumped resistors are usually not very useful at microwave frequencies, thin film resistors deposited directly onto the dielectric substrate are quite satisfactory. In some cases it is possible to connect a very thin wire to the center of a strip- or microstrip line, where very little rf current is flowing, thereby introducing the dc current without disturbing the rf circuit significantly.

The desired high ratio of dc resistance in the bias feed line to rf characteristic impedance can be enhanced by using the circuit in Figure 121. This circuit is best understood by picturing the half-wavelength line as a series pair of quarter-

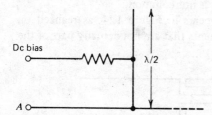

Figure 121 Dc feed circuit.

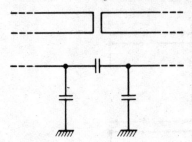

Figure 122 (*a*) Gap in a transmission line. (*b*) Its equivalent model.

wavelength lines. The open-circuited quarter-wave line presents a short circuit (to rf) at the point where the dc series feed resistance is introduced. The parallel combination of resistances, at the signal frequency, is of course zero. This (zero) impedance is reflected through the second quarter-wave line to the signal line, where it presents an infinite impedance at the signal frequency to the rf signal. The useful bandwidth of this circuit is of course dependent on the actual values of the circuit elements involved.

The desired dc isolation of the input rf line can be accomplished simply by providing a narrow gap in the line (Figure 122*a*). Figure 122*b* shows the equivalent circuit of such a gap. The added shunt capacitances are caused by fringing fields at the discontinuities, as discussed in Chapter 10. This shunt capacitance, as well as the series capacitance, may or may not be acceptable for a given application. If a very low VSWR is required, a pair of coupled quarter-wave lines may be used, as shown in Figure 55(3). This circuit is, of course, a bandpass transformer (see Chapter 8). If the lines are very heavily coupled, the effective bandwidth may be so large that the frequency-dependent effects of the coupling circuit are negated. In practice, very heavy coupling between two lines can be achieved in several ways: (1) micro- or stripline center conductors can be simply overlayed with a thin dielectric sheet placed between the layers; (2) the center conductor of a coaxial cable can be made into two concentric layers, again with a thin dielectric sheet placed between the layers; or (3) interdigitated coupling capacitor "fingers" can be used to increase edge coupling (Figure 123). The third technique usually puts severe requirements on the dimensional tolerances of the fabrication technology being used, therefore is not common.

A complete PIN diode SPDT switch appears in Figure 124, as realized on microstrip line. Note that the only components that are not actually part of the

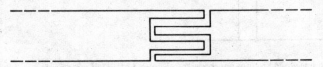

Figure 123 Interdigitated gap in a transmission line.

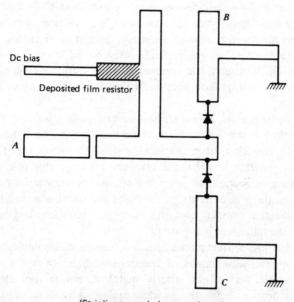

Dc bias

Deposited film resistor

(Stripline ground plane not shown)

Figure 124   Complete SPDT PIN diode switch on a microstrip format.

"printed circuit" are the PIN diodes themselves. This type of circuit is therefore excellent in terms of cost and reliability.

The dc supply and isolation circuit used in the PIN diode switch circuit above is often built as a separate circuit, known as a "bias T." This circuit was shown briefly in Chapter 5. When built as a separate unit, with two rf ports and one dc port, the bias T is a convenient unit for easy connection into an rf circuit for biasing transistors, diodes, and so on. In circuits where critical measurements of dc voltages applied to the devices are necessary, a second "sensing" resistor may be added. Since essentially no current flows through the sensing resistor, the high impedance voltmeter will indicate accurately the actual bias voltage applied to the device.

12.3   ATTENUATOR CIRCUITS

At low frequencies, matched attenuators (pads) are built in the form of T or π networks. As long as the elements are purely resistive, the attenuator will operate independent of frequency. At VHF and UHF T and π attenuators can be carefully assembled and will work well. By cascading different combinations of these units, adjustable attenuators can be made. To build a high frequency variable attenuator, however, devices such as PIN diodes must be resorted to. The frequency response of such an attenuator is a function of the diodes themselves, and also how well the entire assembly—consisting of diodes, dc feeds and blocks, and dc current limiting resistors—may be packaged into a very small space. Still in all, ingenuity has conquered more difficult problems, and PIN diode T and π attenuators have been built that work well up into the gigahertz frequency range.

The T and π attenuators, however, have several basic characteristics that may be objectionable in some cases. First, there must always be at least one diode and one dc blocking element in the signal path. This means that there is a distinct minimum insertion loss that the attenuator will be able to reach that cannot approach zero. Second, to keep the attenuator matched over various attenuation levels, the bias currents to the series and shunt arms must be varied in a fairly complicated manner, and the minimum insertion loss bias does not correspond to the minimum bias current.

At UHF and at higher frequencies, it is practical to realize devices that contain one or more quarter-wavelengths of transmission line. An attenuator circuit is therefore realizable that is not strictly matched, but is very close to being matched under proper circumstances while bypassing by its nature the two basic properties of the T and π attenuators described above. To understand the operation of this attenuator, referred to as a PIN diode absorption attenuator, consider first the circuit in Figure 125. In essence, this circuit consists of a PIN diode shunting a transmission line. The dc bias circuit is represented as lumped elements, for the sake of clarity. Dc blocking is not necessary, since the line is maintained at dc ground (the inductor shown is, again, a lumped element model of the actual dc return circuit). Since there are no series elements in this circuit, the minimum insertion loss is determined solely by the loss of the transmission

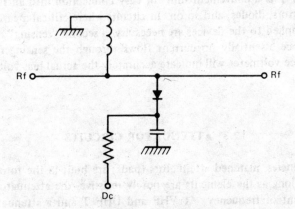

(Transmission line ground returns not shown)

Figure 125   A one-diode PIN absorption attenuator.

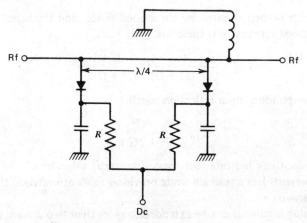

(Transmission line ground returns not shown)

Figure 126 A two-diode PIN absorption attenuator.

line and may be kept very low. Also, a zero bias current condition corresponds to a minimum insertion loss condition. This circuit will attenuate, but is matched only at minimum insertion loss.

Now, consider a circuit (Figure 126) where two PIN diodes are shunting the transmission line, placed a quarter-wavelength apart at some desired operating frequency, Not only will this circuit act as an attenuator, but a mismatch caused by one diode, showing up as a lower-than-Z_0 impedance, is in part counteracted by the second diode, a quarter-wavelength away, where the first mismatch has been "inverted" to a higher-than-Z_0 impedance.

To analyze the operation as just described, assume a transmission line with a characteristic impedance of $Z_0 = 1$. Also, represent the PIN diodes by their effective (bias dependent) rf resistances, $R = 1/G$. Assume also that the line is both driven by and terminated with an impedance Z_0. For the one-diode circuit, the input impedance to the attenuator is

$$Z_{in} = \frac{1}{1 + G} \qquad (12.29)$$

and the reflection coefficient at the input is

$$\Gamma_{in} = \frac{1/(1 + G) - 1}{1/(1 + G) + 1} = \frac{-G}{2 + G} \qquad (12.30)$$

For the two-diode attenuator, the input impedance of the one-diode attenuator (12.29) is reflected down the quarter-wavelength line, becoming

$$Z = G + 1 \qquad (12.31)$$

This impedance is then shunted by the second diode, and the input impedance for the two-diode attenuator is therefore

$$Z = \frac{1}{G+1(1+G)} = \frac{G+1}{G^2+G+1} \qquad (12.32)$$

with the corresponding input reflection coefficient,

$$\Gamma_{in} = \frac{-G^2}{G^2+2G+2} \qquad (12.33)$$

As these equations indicate, for every (nonzero) value of G, the two-diode attenuator presents less mismatch while providing more attenuation than its one-diode counterpart.

The foregoing argument can be extended to more than two diodes. Also, it has been shown that by biasing the diodes with fixed but unequal ratios of current, the mismatch over some attenuation range can be optimized. The results over a large dynamic and frequency range can be made arbitrarily good by specifying a calculable number of diodes. This design procedure has been optimized, and is referenced at the end of this chapter.

12.4 SUGGESTED FURTHER READING

1. L. Dworsky, "Computer-Optimized Design of PIN Diode Absorption Attenuators," *Microwave Journal*, Vol. 19, No. 2, February 1976. The absorption attenuator is designed optimally for various dynamic attenuation ranges using from one to six PIN diodes.

2. H. Watson, *Microwave Semiconductor Devices and Their Circuit Applications*, McGraw-Hill, New York, 1969. Although this is principally a semiconductor book, many circuits are realized using transmission line networks, and the interplay is well presented.

Index

CHEMISTRY-CHEMICAL ENGINEEING

PHYSICS

6

7

ELECTRICAL-ELECTRONIC ENGINEERING

9

MECHANICAL ENGINEERING

S-10-1 Shtiplman : DESIGN AND MANFACTURE OF HYPOID GEARS $190
S-10-2 Spalding : COMBUSTION & MASS TRANSFER $300
T-10-1 Towill : COEFFICIENT PLANE MODELS FOR CONTROL SYSTEM AN
                DESIGN (1981)...................................$190
T-10-2 Tanner : INDUS RIAL ROBOTS - Applications - 2/e 1981......$400
W-10-1 Wilson : HANDI JOK OF FIXTURE DESIGN $200
W-10-2 Weller : MACHINING FUNDAMENTALS : A Basic Approach to
                Metal Cutting & Exhibit, Test (1980) $400
W-10-1 Wilson : HANDBOOK OF FIXTURE DESIGN $200
W-10-2 Weller : MACHINING FUNDAMENTALS : A Basic Approach to
                Metal Cutting & Exhibit, Test (1980) $400

MATERIAL

A-11-2 Aronson : DIFFUSION ....$
B-11-1 Bube : PHOTOCONDUCTIVITY OF SOLIDS $250
B-11-2 Bond : CRYSTAL TECHNOLOGY $230
B-11-3 Berry : THIN FILM TECHNOLOGY...............................$390
B-11-4 Briggs : PRACTICAL SURFACE ANALYSIS-by Auger & X-ray
                Photoelectron Spectroscopy (1983) $360
● B-11-5 Bergeron : INTRODUCTION TO PHASE EQUILIBRIA IN CERAMICS
                (1984) $270
● B-11-6 Bunget : PHYSICS OF SOLID DIELECTRICS (1984) $300
C-11-1 Czanderna : METHODS OF SURFACE ANALYSIS $300
C-11-2 Chu : BACKSCATTERING SPECTROMETRY $260
C-11-3 Cowley : DIFFRACTION PHYSICS 2/e (1981) $290
C-11-4 Cottrell: AN INTRODUCTION TO METALLURGY 2/e..............$360
C-11-5 Cottrell : DISLOCATIONS & PLASTIC FLOW IN CRYSTALS.......$200
● C-11-7 Chapman : QUANTITATIVE ELECTRON MICROSCOPY (1984) $350
E-11-1 Evans : AN INTRODUCTION TO METALLIC CORROSION (1981) $250
F-11-1 Faktor : GROWTH OF CRYSTALS FROM THE VAPOUR $140
■ F-11-2 Inerath: CERAMIC MICROSTRUCTURES '76....................$660
F-11-3 Fine : INTRODUCTION TO PHASE TRANSFORMATIONS IN CONDENSED
                SYSTEMS ....$ 70
* F-11-4 Feidman: MATERIALS ANALYSIS BY ION CHANNELING(1982) $250
G-11-1 Geiger : TRANSPORT PHENOMENA IN METALLURGY...............$360
G-11-2 Gordon : THE NEW SCIENCE OF STORONG MATERIALS (1982) $200
G-11-3 Gaskell : INTRODUCTION TO METALLURGICAL THERMODYNAMICS 2/e $350
G-11-4 Gabe : PRINCIPLES OF METAL SURFACE TREATMENT &
                PROTECTION 2/e $150
G-11-5 Guy : ESSENTIALS OF MATERIALS SCIENCE.................... $320
● G-11-6 Guinier : THE STRUCTURE OF MATTER (1984) $220
● G-11-7 German : POWDER METALLURGY SCIENCE (1984) $200
H-11-1 Hatch : ALUMINUM-Properties & Physical Metallurgy (1984).$360
● H-11-2 Hench : ULTRASTRUCTURE PROCESSING OF CERAMICS CLASSES
                & COMPOSITES (1984) $380
K-11-1 Kohl : HANDBOOK OF MATERIALS AND TECHNIQUES FOR VACUUM
                DEVICES $400
K-11-2 Kazmerski : POLYCRSTALLINE AND AMORPHOUS THIN FILMS AND
                DEVICES $160
K-11-3 Kroger : THE CHEMISTRY OF IMPERFECT CRYSTALS(Vol.1-3)....$900
K-11-4 Kiddle : MATERIALS TECHNOLOGY FOR TEC (Level II)1983 .....$120
K-11-5 Khachaturyan : THEORY OF STRUCTURAL TRASFORMATIONS IN
                SOLIDS(1983).................................... $360
● K-11-6 Klar : POWDER METALLURGY - Applications Advantages &
                Limitations (1983) $200
● K-11-7 Kuczynski : SINTERING & HETEROGENEOUS CATALYSIS (1983) ..$250
L-11-1 Lenel : POWDER METALLURGY-Principles and Applications(1980).$360
L-11-2 Leslie : THE PHYSICAL METALLURGY OF STEELS (1981) $290
L-11-3 Levinson : GRAIN BOUNDARY PHENOMENA. ELECTRONIC
                CERAMICS (1981) $350
L-11-4 Luborsky: AMORPHOUS METALLIC ALLOYS(1983)................$380
M-11-1 Maissel : HANDBOOK OF THIN FILM TECHNOLOGY ....$

12

FOOD,MANAGEMENT,OTHER

OPTICAL ENGINEERING & ELECTRO-OPTICS

化學，食品，化工，物理，其他

16